Haim Shapira was born in Lithuania in 1962. In 1977 he emigrated to Israel, where he earned a PhD in mathematical genetics for his dissertation on Game Theory and another PhD for his research on the mathematical and philosophical approaches to infinity. He now teaches mathematics, psychology, philosophy and literature. He is an author of nine bestselling books. His stated mission as a writer is not to try to make his readers agree with him, but simply to encourage them to enjoy thinking. One of Israel's most popular and sought-after speakers, he lectures on creativity and strategic thinking, existential philosophy and philosophy in children's literature, happiness and optimism, nonsense and insanity, imagination and the meaning of meaning, as well as friendship and love. He is also an accomplished pianist and an avid collector of anything beautiful.

BY THE SAME AUTHOR:

Conversations on Game Theory
Things That Matter
The Wisdom of King Solomon: A Contemporary Exploration of Ecclesiastes and the Meaning of Life
Gladiators, Pirates and Games of Trust: How Game Theory, Strategy and Probability Rule Our Lives
Nocturnal Musings
A Book of Love
Happiness and Other Small Things of Absolute Importance
A Most Beautiful Childhood Memory

EIGHT LESSONS ON INFINITY

A Mathematical Adventure

Haim Shapira, PhD

Translated from the Hebrew

by Linda Yechiel

DUNCAN BAIRD PUBLISHERS

LONDON

To Daniela, Tal and Inbal

This edition first published in the UK and USA 2019 by
Duncan Baird Publishers, an imprint of Watkins Media Limited

Unit 11, Shepperton House
89–93 Shepperton Road
London
N1 3DF

enquiries@watkinspublishing.com

1 3 5 7 9 10 8 6 4 2

Typeset by Integra Software Services Pvt. Ltd, Pondicherry

Printed and bound in the United Kingdom
by TJ International Ltd, Padstow, Cornwall

A CIP record for this book is available from the British Library

ISBN: 978-1-78678-184-0

www.watkinspublishing.com

Contents

ACKNOWLEDGEMENTS

First and foremost, I would like to thank Etan Ilfeld for having confidence in me and my books.

I'd like to acknowledge with gratitude my faithful translator Linda Yechiel.

I would like to express my special thanks to Alain Dekker, who never stopped challenging me, for his enormous help and patience.

To Tom Benhamou, a set theory expert, who wisely edited my book and is responsible for many great insights – a very big thank you.

I would like also to thank thc book's project manager, Slav Todorov, and to express my appreciation for everyone at Watkins who laboured over this book.

Last but never least, I would like to thank my agents, Vicki Satlow and Ziv Lewis.

INTRODUCTION

If I were again beginning my studies, I would follow the advice of Plato and start with mathematics.

(Galileo Galilei)

The English biologist and public intellectual Richard Dawkins once noted that no one is ever proud to admit that he or she is a boor and an ignoramus with respect to literature, but it is socially acceptable to proudly admit total illiteracy when it comes to science, with the epitome being perfect illiteracy with respect to mathematics. Dawkins was not the first to note this – he himself points out that this concept has long become a cliché.

It is very true, of course. No one would ever boast that they have never read a book, never seen a work of art, or never – not even once – been moved by a piece of music. If we were to conduct a survey, I am pretty sure we would discover that no educated adult has never heard of Shakespeare, Rembrandt, or Bach. They also probably recognize the names of the great mathematicians Pythagoras, Isaac Newton, and Albert Einstein. But how many have heard of Leonhard Euler or Srinivasa Ramanujan or Georg Cantor?

Perhaps, at this very moment, you are also asking yourself, "What? Who are these people? Those names are unfamiliar to me."

These people are great mathematicians. Very great mathematicians!

I am a serious music, literature and art enthusiast, but I honestly believe that Ramanujan's mathematical formulas are no less marvellous than Bach's musical structures, and that Cantor's discoveries about infinity are as wondrous to me as Shakespeare's works.

And if we're already comparing cultural genius with mathematical genius, I would like to point out that Cantor was an expert on Shakespeare, and Einstein was an accomplished pianist and violinist. This is a very common phenomenon, and I know of many mathematicians who are extremely knowledgeable in literature, art, and music. In fact, German mathematician Karl Weierstrass once said that a mathematician who is not a bit of a poet will never be a good mathematician. Yet, it seems the road does not go both ways: not a few people from the field of literature, music, or the arts seem to have an aversion to mathematics.

Why is this so? Why do so many people, as educated as they may be, shy away from the intricacies and beauty – yes beauty – that can be found in the world of numbers and in their relationships one to the other?

Perhaps the main reason for this is mathematics' inapproachability and the difficulties it places before those who wish to know her. It is true that mathematics is quite complicated, and one needs to devote time and thought to understanding its intricacies, but sometimes one has to dive to the deepest depths to find an especially exquisite pearl.

The idea to write this book came to me one day as I was perusing my collection of books on mathematics. I noticed that the majority of my books fell into one of two categories:

1 Mathematics books written for the lay audience. Some of these are wonderful books, but they focused on stories about mathematics and not on the mathematics itself.
2 Mathematics books written for mathematicians. In this category, too, are many excellent books, but only mathematicians are able to read (and understand) them.

So, I decided to write a book that would be a "third category". I wanted to introduce to a lay audience, simply and clearly, two mathematical theories that I consider the most intriguing – number theory and set theory, both dealing with infinity. Alongside, I would offer some strategies in mathematical thinking that would allow the reader to test his ability in solving some truly fascinating mathematical problems.

It is important to me that anybody who is curious and likes to put on a thinking cap once in a while will enjoy this book, so I restricted myself from using any fearsome mathematical symbols (you won't find any $\nabla f\left(x_1,\ldots,x_n\right), \frac{\partial f}{\partial x}, \iiint, \lim_{x\to\infty}$ anywhere in this book). Only the most basic mathematical operations will be used (addition, subtraction, multiplication, and division, plus a couple of more "advanced" terms such as powers and roots), but that does not mean that you won't be required to engage in some deep thinking. I also did my best to keep the text light and give it an amusing lilt – nobody really likes questions about three ducks who are filling a pool while two ducks (for reasons known only to them) are simultaneously trying to empty it.

Comments on the book, solutions to the questions, and questions about the questions can be sent to shapirapiano@gmail.com. Wishing you a fascinating journey.

WARM-UP

A SHORT INTRODUCTION TO THINKING

Thinking: the talking of the soul with itself.

(Plato)

If you bothered to read the introduction above (why is it that so many people never bother reading introductions?), you already know that I have a sizeable collection of mathematics books. One of my favourite occupations is to mess around with interesting problems. Well, naturally, *I* would. That's what I studied. But you don't have to have be at university level to see and enjoy the exquisite beauty of mathematics. If you have the patience to put on your thinking cap for a little while, there are thousands of interesting – some quite famous – mathematical problems and paradoxes that have delighted young and old for centuries. With just a little effort, almost anyone can experience the elation of being able to solve something that at first seems quite complex.

In this section, I present a modest assortment of some of my favourite mathematical problems, from the somewhat simple, through to the pretty profound, and all the way to unquestionably unsolvable (and if you do solve it, there is a prize waiting for you). The point is to give you, dear reader, a tiny taste of the great variety of interesting cogitation that is yours to discover in the amazing world of mathematics.

Great little study – an open problem

Many years ago, I read the Pulitzer Prize-winning *Gödel, Escher, Bach* by Douglas R Hofstadter. The author himself describes the book as "a metaphorical fugue on minds and machines in the spirit of Lewis Carroll". It spoke about a wide variety of subjects in the realms of mathematics, music, symmetry, artificial intelligence, and logic, and included a host of mathematical riddles. I would like to present you with one of them.

Think of any number, and by number I mean a whole or counting number. (Achilles, that hero with the problematic heel, and also one of the characters in Hofstadter's book, thought of 15. You can, of course, choose any number you like.)

Now, proceed as follows: if the number is even, divide it by 2. If it is odd, multiply it by 3 and add 1. Repeat this procedure over and over until you arrive (if you do) at the number 1. Let's see how it works.

15 is an odd number so we multiply by 3 and add 1.
$15 \times 3 + 1$ results in 46.
46 is even, so we halve it and get 23. This is odd, so we multiply by 3 and add 1.
$23 \times 3 + 1 = 70$
We continue the process:
$70/2 = 35$
$35 \times 3 + 1 = 106$
$106/2 = 53$
$53 \times 3 + 1 = 160$

160/2 = 80
80/2 = 40
40/2 = 20
20/2 = 10
10/2 = 5
5 × 3 + 1 = 16
16/2 = 8
8/2 = 4
4/2 = 2 and finally 2/2 = 1.
The process has reached its end.

The question is, will the process eventually lead to 1 for *any* number at all?

Why don't you try a couple of other numbers? Depending on the number, the process may end up being extremely long and you may need a very big piece of paper. If you try to run the process on a computer, be forewarned – the computer may work overtime.

Hofstadter suggested to Achilles to try 27. You may like to do the same. I'll give you a couple of minutes … maybe hours.

Did you give up? If you start with 27, the process seems to go on and on into an endless string of calculations that never ends. You might get to the point where you believe it will never end. In fact, the number of steps needed is 111.

In his book, Hofstadter warns Achilles about trying to find an answer to the question above (does every number lead to 1?), and reveals that this is known as the "Collatz conjecture". The Collatz conjecture is one of a great many "open problems" in mathematics. Open problems are mathematical questions that nobody has yet been able to solve. The Collatz conjecture ("conjecture", just in case you're not sure, is another word for "speculation", or, to be more exact, "a proposal for a new possible theorem which needs to be proven") proposes that, no matter which number you begin with, the process described above will lead eventually to the value of 1. The conjecture is named after German mathematician Lothar Collatz (1910–1990), who first presented it in 1937. Nevertheless, this conjecture has at times been called, among other names, the Ulam conjecture (after Polish mathematician Stanisław Ulam) or Kakutani's problem (after Japanese mathematician Shizuo Kakutani). Sometimes it is simply known as the $3n + 1$ conjecture, which makes a lot of sense.

When I first met the $3n + 1$ conjecture, I was too young to appreciate how very difficult and profound it was. I assumed that it wouldn't take more than a few days for me to come up with the criteria defining which numbers will yield 1 as the final step. In fact, I figured I could prove that the conjecture was true and that all numbers eventually do lead to 1. Perhaps, while I was at it, I could even discover the distribution of the number of steps required for each particular number (for example, there were 17 steps when we tried 15). The only thing that really dumbfounded me was how it was possible that no one else had yet successfully done it.

Or so I thought …

There is, it seems, a good reason that this is considered an "open problem".

Even though I was unsuccessful in my mission, I was not too upset. I find a lot of appeal in difficult questions. They force one to think. In fact, I prefer questions that I can't solve (or at least not solve too easily) over those that I can solve in a jiffy and without much intellectual effort. Of course, this doesn't mean that my epitome of delight is when I am unsuccessful in solving some problem – undoubtedly, there is much greater pleasure in the solution of a difficult problem after much toil.

But let us return to our conjecture. Note what is happening here. We have encountered a mathematical problem that uses only basic arithmetical operations – addition, multiplication and division – yet *nobody in the world knows how to solve it!*

How can this be possible? One would think that a problem that can be stated so simply would be a problem with a simple solution. Well – no! A simple question doesn't always have a simple answer. There are many questions in mathematics that one can ask small children, and they will easily understand the problem, yet even the most brilliant adults have not thus far discovered the solution.

After examining a good number of cases of our Collatz problem, a phenomenon will become apparent, namely that the last numbers appearing in the process are consistently decreasing powers of 2. For example, if we start from 15, the last five numbers in the series are 16, 8, 4, 2, and, finally, 1.

To express this phenomenon as a rule, we can state that if at any point we happen to reach a number 2^n, from there the path to 1 is guaranteed by dividing by 2 exactly n times. This observation suggests a rephrasing of the $3n + 1$ conjecture: no matter what number we start on, will we, at some point, reach a power of 2?

The principle of replacing a given problem with another problem is called reduction. This method is a useful mathematical tool and, in a certain sense, provides a more natural method for solving mathematical problems. Another, similar, problem-solving strategy is that of "backward thinking" (from finish to start) and may be familiar to some of you from the world of mazes. Sometimes, when searching for a way out of a maze, it is more productive to begin at the maze's exit and work one's way to the starting point. In some profound sense, this is exactly what we are doing when we use mathematical reduction.

Hungarian mathematician Paul Erdős (1913–1996) enjoyed offering cash prizes to anyone who successfully solved open problems in mathematics that interested him. The prizes started at $25, and the Collatz conjecture was worth, according to his price list, $500 – an amount that meant it was in the category of pretty expensive questions, even though Erdős himself said that the world of mathematics may not be ready for questions as hard and confusing as the $3n + 1$ conjecture. Erdős has since left this world, but don't worry, his colleague Ron Graham has taken over the job of paying out prizes. If you do happen to solve this problem, you can get the prize in one of two ways: either a check signed by Erdős himself before his death (meant for framing only; it has passed its "cash by" date) or as a cashable check (a struggle between the sin of pride and the sin of greed).

(As an aside, and because it is an interesting fact that I'd like to pass on, the largest number ever used in a mathematical proof is named after this very Ron Graham. The number is so large that it cannot be written using standard mathematical notation.)

Wisdom is knowing that you don't know what you don't know and that you do know what you know. Stupidity is thinking that you know things you don't know or that you don't know things you actually do know.

(Chinese proverb)

ERDŐS NUMBER

Paul Erdős was an exceptionally prolific mathematician. (A wonderful biography about him is *The Man Who Loved Only Numbers* by Paul Hoffman.) Erdős wrote more than 1,400 academic papers. He was an enthusiastic advocate of teamwork and cooperation, and a grand total of no less than 511 mathematicians collaborated with him in writing his papers over the years. Any mathematician who wrote a paper with Erdős himself was awarded the prestigious title "Erdős 1". Those who collaborated with "Erdős 1" individuals but never actually with Erdős himself were given the title "Erdős 2". Similarly, the titles Erdős 3, 4, and so on were conferred on others down the line. The rule was: If you collaborate on a paper with someone whose lowest Erdős number is k, your Erdős number is k + 1. Erdős himself was the one and only "Erdős 0". At the other end of the spectrum, someone who never wrote a paper with Erdős and also never wrote a paper with anyone with a finite Erdos number, received the number Erdős infinity (Erdős ∞). "Erdős infinity" may sound pretty impressive – perhaps even more impressive than "Erdős 7", but not a few of you will certainly be surprised to discover that you (like most of humanity) bear the title "Erdős ∞". I, myself, don't write papers, but I once collaborated on a paper with an "Erdős 3" mathematician, and, therefore, without even aspiring to it, I am the proud bearer of the title "Erdős 4".

This brings me to a popular parlour game called "Six Degrees of Kevin Bacon". Kevin Bacon, a famous

Hollywood actor, once claimed that every single actor in Hollywood has either worked with him (Bacon 1) or has worked with someone with whom he has worked (Bacon 2), or with someone who has worked with someone … (Bacon 3, 4 …) The claim is that almost every actor or actress in Hollywood has a "Bacon number" not greater than six. Elvis Presley, for example, is a Bacon 2. I leave it up to you to discover the connection.[1] The world, it seems, really is a small place, because there are some people who have both an Erdős number and a Bacon number. Ron Graham, for example, is "Erdős 1" and "Bacon 2". The famous Israeli actress Natalie Portman is "Erdős 5" and "Bacon 1" (surprised, aren't you?).

And now we return to the solution of the Collatz conjecture. Well, there isn't one, and the truth is I know a lot of ways to earn five hundred dollars that are easier than fooling around with the Collatz conjecture. But who am I to tell *you* what to do?

A chequerboard riddle

I deliberated over whether to present the next riddle. It is really very simple. Nevertheless, after a rousing discussion between me and myself, I decided to present it, as it is quite famous, and both the riddle and the solution are remarkably charming.

Let us examine an 8×8 grid.

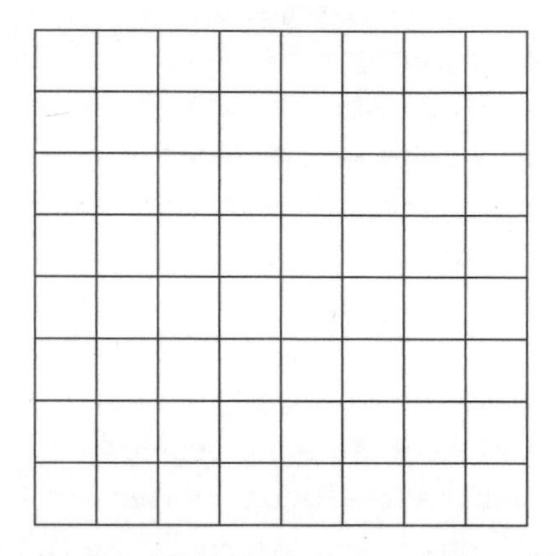

It is clear that the grid can easily be covered with 32 domino pieces that are 1×2. Now, let us remove one square from two opposite corners.

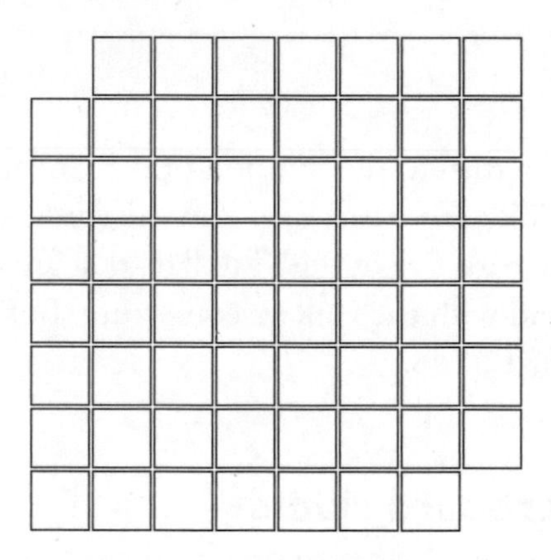

Will it be possible to cover the grid with only 31 dominoes?

Most of my friends to whom I present this riddle (none of whom are mathematicians but most of whom are quite clever) were sure that the answer would be "yes" – one need only to put one's mind to how to figure it out.

But the correct answer is "no". No matter what you do, it will be impossible to cover a grid missing two opposite corners with 31 domino pieces.

If we convert the all-white grid into a chequerboard, the reason this is so becomes almost immediately apparent.

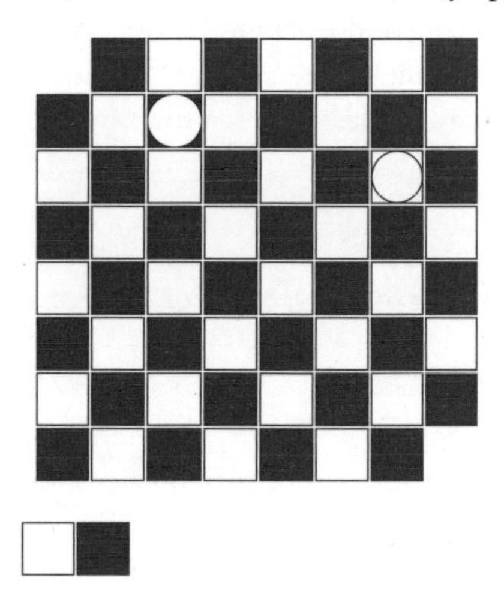

As you can see, each domino piece can cover one black square and one white one; therefore 31 dominoes can cover precisely 31 white squares and 31 black ones. Since the two squares removed from the grid are of the same colour – white – our concatenated board has 30 white squares and 32 black squares. Therefore, this board can never be covered with 31 dominoes.

Many years ago, I was a student of mathematics at Tel Aviv University, and I taught a course for "science-oriented youth" called "Paradoxes, Riddles, and Numbers". I would present the above riddle to the young scholars who had signed up for this course. A curious thing always happened. Many students refused to accept the proof that showed that it was impossible to cover a board with two opposite corners removed with 31 dominoes. Interestingly, these included students who, on the face of it, comprehended the explanation perfectly; nevertheless, they would stubbornly arrange and rearrange those dominoes, trying to cover that missing-corners board. I didn't try to dissuade them from their hopeless task – a man must learn from his errors.

> *History teaches us that men and nations behave wisely once they have exhausted all other alternatives.*
>
> (Abba Eban)

Brain-twister

Prove that if two squares of *different* colours are removed anywhere from a chequerboard, it will *always* be possible to cover all the squares with 31 domino pieces.

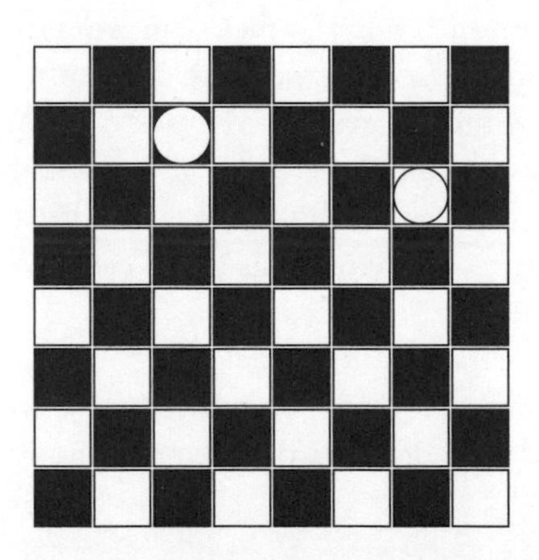

The infinite game of Tic-tac-toe

When I was in elementary school in Vilnius, Lithuania, where I was born, one very significant achievement of mine was acquiring proficient skill at playing pen-and-pencil strategy games during class without any of my teachers catching on. My favourite one was an infinite version of Tic-tac-toe, also known as noughts and crosses. More than once, this game helped me survive the boredom of some of the classes that I was forced to attend.

I will explain the rules of the game.

You are, no doubt, familiar with the regular format of Tic-tac-toe played on a 3 × 3 grid. This is a game meant for kids up to the age of six. Beyond this age, the game should always end in a draw, unless one of the players dozes off during the game. (Which could certainly happen, as the game is really pretty boring.)

In the infinite version, the game is played on an infinite grid, and the goal is to assemble a row of five Xs or five Os. Similar to the original game, the row can be horizontal, vertical, or diagonal. Each player writes an X or an O in succession, and the first one who succeeds in accumulating a row of five is declared the winner.

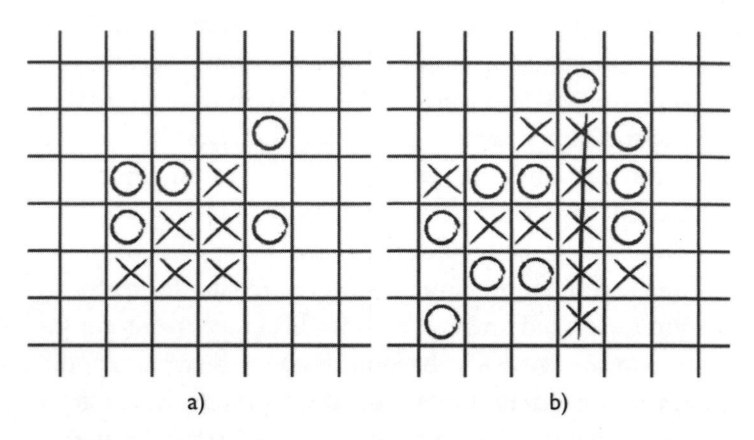

a) The O player has no move to counter the two "open" triple Xs and is about to lose.

b) An example of another game that the X player has just won.

When I "discovered" this game in elementary school, I thought I'd invented it, but later I found that this was not so, and that there is a game called "Gomoku" which is very similar to infinite Xs and Os. Gomoku is especially popular in Japan and Vietnam. "Go" means five in Japanese.

You have likely heard of the game Go. However, even though Gomoku is often played on a board identical to the one used for that renowned epic game, there is no connection between the two. Go is an ancient Chinese game that Confucius even mentioned in his *Analects*. Because it reached the West by way of Japan, it is known by its Japanese name, but as I said, Go is not Gomoku.[2]

Despite the experience I accrued from countless games during classes – and sometimes during recess (it wasn't as much fun playing during recess because it was allowed!) – I couldn't figure out whether the first player (X) can always win given the correct strategy and no matter how the opponent plays, or whether the game will always end in a draw if both opponents play correctly (or, to be more exact, they will never be able to finish the game). My intuition told me that there was some strategy that would assure a win for the opening player.

To be fair, I must admit that it has been decades since I've played this game. I reminded myself about it while writing this book. But my curiosity about the strategic perspective of the game and whether there is some strategy for winning still stands. I am even willing to wager that there is, indeed, some winning strategy. When I'm older and have more free time, I intend to attempt to discover what it might be, but since my plans for this are far into the distant future, you are invited to try to do it before I get to it and save me the effort.

The monk and the riddle[3] – Looking at it from both sides

Early one morning, as the sun was rising, an old Buddhist monk started his climb up a steep and winding mountain toward the monastery on the summit. The old monk ascended along a narrow, twisting path – the only path leading to the monastery. The effort was exhausting.

He proceeded at varying speeds, resting a bit here and there, murmuring mantras, and sometimes stopping to eat a morsel of food or drink some water. He arrived at the monastery on the summit exactly as the sun began to set. The monk spent a few days in the monastery teaching the young monks about compassion, about the Four Noble Truths, about *sunyata* (emptiness), about the illusion of the ego, about *samsara* and suffering, about karma and calmness, about the Noble Eightfold Path, about the teachings of Nagarjuna and about the desire not to have desires.

Once he had finished his teaching, it was time for the monk to descend the mountain and return to his village. He began his journey down, just as he had his climb up, when the first rays of sun appeared and followed the exact same path as before. The old monk's rate of descent was, of course, much faster than his rate of ascent. When he reached the bottom, the monk thought to himself that there was surely some point on the path that he had tread on at the exact time of day both when ascending and descending.

Brain-twister

How did the monk arrive at this conclusion? If you need to think about this for more than ten seconds without solving it, here is a pretty obvious hint:

> Two monks set out at sunrise, one at the foot of a mountain, going up, and the other at the top, going down. They will definitely meet at some point.

The mathematics of tennis – How much is infinity?

Version one

In 1953, English mathematician John E Littlewood (1885–1977) presented the following paradox, known today as the Ross–Littlewood paradox.

An infinite row of tennis balls, numbered 1, 2, 3, 4 …, are lined up at the entrance of a huge, empty room. The hour of midnight is fast approaching. Thirty seconds before 0:00, balls 1 and 2 are introduced into the room, and ball number 1 is immediately ejected. Fifteen seconds (a quarter of a minute) before 0:00, balls 3 and 4 are placed in the room, and ball number 2 is removed. An eighth of a minute before 0:00, balls 5 and 6 are placed into the room, and ball number 3 is removed, and so on and so forth. (In "mathspeak" we would say: at $\left(\frac{1}{2}\right)^n$ minutes before 0:00, balls $2n-1$ and $2n$ are introduced, and ball n is ejected.)

The question is, how many balls will be in the room at exactly 0:00?

Those who deal with this question point out that there are two possible answers, the popularity of each one being almost identical: an infinite quantity and none. How is that possible? Let us examine the logic involved in each.

Infinite: At the end of the process, there will be infinite balls in the room because at each stage of the infinite stages one ball is added into the room (two balls are introduced, but one is removed). Mathematicians say it as follows: for every n, a precise time can be found when the number of balls is $n + 1$. Therefore, there will be an infinite number of balls at 0:00.

None: There will not be a single ball at 0:00 because it is possible to give the precise time when each ball is removed from the room. Ball 1 is removed when the clock shows half a minute to midnight, ball 2 exits when the clock's hands are at a quarter of a minute to midnight, and so on. Translated into more mathematical language, ball *n* exits the room at exactly 1/2 in the power of *n* of a minute before midnight.

If a survey would be conducted, which answer would you vote for?

What is important to realize here – and maybe a little hard to get your head around – is that there are infinite points of time until midnight, since it is always possible to halve the time period remaining.

I would go with the first answer as the "correct" one, and I would venture to say that those who side with the second answer probably cannot break out of a finite thinking schema. Their need to know which balls will be in the room "at the end" of the process is actually similar to a need to know which numbers are "at the end" of the sequence of natural numbers, that is to say, "at the end" of 1, 2, 3, 4, 5, 6, 7, 8, 9, …, 12,367, 12,368, …, …, …

We all know and understand that the set of natural numbers is infinite, but nobody and no one can possibly say which numbers will be "at the end" of this series, simply because there is no end.

It is interesting to note that Saint Augustine (354–430) believed that God saw and knew all the infinity of the natural numbers and all their properties and by doing this converted them in some way to a finite set (that was Saint Augustine's viewpoint, of course).

Here are two more variations of the Ross–Littlewood paradox.

Version two

Again, we have before us an infinite row of tennis balls numbered 1, 2, 3, 4 ... lined up at the entrance to a huge, empty room. Half a minute before midnight, balls 1, 2, 3, 4, 5, 6, 7, 8, 9, and 10 are introduced into the room, and ball number 1 is tossed out. A quarter of a minute before midnight, balls 11, 12, 13, 14, 15, 16, 17, 18, 19, and 20 are introduced, and ball number 2 is ejected, and so on.

Same question, of course: how many balls will be in the room at exactly midnight?

Here, we are introducing 10 balls and removing only one at each stage, that is to say, a net addition of nine balls. Since we are repeating the process infinite times, it seems absolutely clear that at midnight there will infinite balls in the room. (You could even say that it will even nine times infinity!)

Brain-twister

Can you tell me *which* balls will be in the room? By which I mean the number(s) on the balls in the room.

Third version

Same row of tennis balls numbered 1, 2, 3, 4 ..., and still all lined up outside that huge, empty room. Half a minute before 0:00, balls 1 and 2 are introduced into the room, and ball number 2 is tossed out. A quarter of a minute before midnight, balls 3 and 4 go in, and number 4 goes back out. And so forth. Same question: how many balls will be in the room at midnight?

Suddenly everything becomes crystal clear.

Because we are tossing out all the balls with even numbers, at midnight there will be an infinite number of balls in the room, all with odd numbers. So these are the occupants of the room at midnight: 1, 3, 5, 7, 9, 11, 13, 15 …

Of course, there are infinite odd numbers, and they will all be in the room. The even numbers also form an infinite set, but they will be outside.

Another brain-twister

Are the sets of odd numbers and even numbers smaller than the set of *all* the natural numbers? (A natural number is a positive integer.)

At first, you might assume that this must be so. It seems to make sense that the set of even numbers, for example, would be only half as large as the set of natural numbers (which includes both odd and even).

However, let's look at it this way: every even number can be matched to any natural number.

1	2	3	4	…
2	4	6	8	…

Now we can appreciate the mind-boggling notion that even though the set of even numbers skips every other number (compared to the set of natural numbers), both sets still have exactly the same amount of elements. We say they have the same cardinality. We will be talking much more about the concept of cardinality later on in this book.

This essentially brings us to an even more profound question: is it even possible to compare infinite sets of numbers and ask which set is larger? Do the words "greater" and "fewer" or "larger" and "smaller" actually have any significance when talking about infinite magnitudes?

Keep on reading!

The concept of infinity is a complex and profound one, and truly is sometimes mind-boggling. It is worth noting Galileo's notion on the subject:

> *The difficulties in the study of the infinite arise because we attempt, with our finite minds, to discuss the infinite, assigning to it those properties which we give to the finite and limited; but this … is wrong, for we cannot speak of infinite quantities as being the one greater or less than or equal to another.*
>
> (Galileo Galilei)

Despite my affection and appreciation of Galileo Galilei, I am more hopeful. In the remainder of this book, we shall be dealing quite a bit with infinity, even though we are, unfortunately, painfully finite creatures. As Pascal said:

> *Man is only a reed, the weakest in nature, but he is a thinking reed.*
>
> (Blaise Pascal)

Bouncing it around one more time

If you are still not convinced that (in all the versions) there will be an infinite number of balls in the room at midnight, I will now release my doomsday weapon and put forward this final version: imagine that the balls are not numbered; that they are all just regular white tennis balls.

Numbering or not numbering the balls should not affect the quantity of the balls at midnight.

Now it all should be crystal clear. If at every stage the net number of balls increases and if there are infinite stages till 0:00, at midnight we will have an infinite number of balls.

Now we can answer the question of *which* balls will be in the room.

There will be infinite … white balls![4]

This last version is unique compared to the former ones since there is no rule to indicate which particular balls are thrown out of the room. When the balls are numbered, we have the privilege of being able to suggest a rule. However, now all the balls are identical, we are forced to randomly determine which will be ejected.

April Fool's Day, or "Logic in Big Brother's House"

Renowned logician, magician, and mathematician Raymond Smullyan (1919–2017) (he was also a concert pianist, by the way – you can listen to him playing Bach on YouTube) discussed how he first met the concept of logic. It happened one April 1st, when Raymond was a little boy. The night before, the future logician's older brother promised that he would play a trick on him (which is what one usually does on the first of April), and assured him that no matter how much Raymond would try to avoid being tricked, he wouldn't be able to.

Raymond took the threat very seriously and decided that, by hook or by crook, he would not give his brother the pleasure of tricking him. After thinking about it for a little while, young Raymond decided that the best way to avoid an April Fool's Day prank would be to hole himself up in his room and not come out for the entire day.

Clever, no?

Raymond went into his room, closed the door and sat, totally bored, hour after hour after hour … until midnight, at which point he proudly came out of the room, bragging to his brother that his plan to trick him had failed. His brother answered, "Wrong! I tricked you! You expected me to trick you, but in the end, I didn't trick you, and so I tricked you! Ha!"

Until his dying day, Raymond Smullyan was not sure whether his brother had really managed to trick him or not. What do you think?

Chocolate and poison

This is quite a simple game, better known as Chomp. (The chocolate-bar formulation of Chomp is due to the late American mathematician David Gale, and the name Chomp was invented by Martin Gardner.) It is played on a chequered board and these are the rules:

The opening player marks one of the boxes X.

		X		
Poison				

Once that is done, all the boxes to the right and above the X also turn X (and are eliminated). The original X is in bold:

		X	X	X
		X	X	X
		X	X	X
Poison				

Now it is the other player's turn to mark an empty remaining box with O. Once that happens, all empty boxes to the right and above it are marked O too (original O in bold):

		×	×	×
		×	×	×
		X	×	×
			○	○
Poison			**O**	○

Then the first player marks another X and the second player marks another O until one of them is forced to choose the poison and dies (metaphorically, of course).

Warning: This game is addictive!

You are welcome to try playing this on a 7 × 4 board (7 rows and 4 columns or vice versa).

If the game is played on a board that has an equal number of rows and columns, there is a strategy by which the opening player will always win. Can you find it? Take three minutes to think.

Solution

The first player must choose the box right above the poison, diagonally:

	×	×	×	×
	×	×	×	×
	×	×	×	×
	X	×	×	×
Poison				

From now on, the first player just has to follow symmetrically:

O	X	X	X	X
O*	X	X	X	X
	X	X	X	X
	X	X	X	X
Poison			**X**†	X

* Rival's first choice
† First player's response

How this one is won should be clear now.

Things become much more complicated when this is played on a board where the number of rows and columns are not equal, but we can still prove that the opening player has a winning strategy. Unfortunately the proof does not specify any particular strategy. Mathematicians call such proofs "non-constructive proofs of existence".

Let's finish this lesson with an exercise:

Find the winning strategy for the first player when this game is played on 2 × N rectangle (2 rows, N columns).

Hint: In order to get symmetry, as in the square case, we have to get to the point where only the poison square is left or there is one square open above the poison and one to the right.

Now, after solving this problem (I hope so), what do you think will happen in the case of two rows and an infinite number of columns? Who will win now? *Infinity is a great law-breaker!*

LESSON 1

THE WONDERFUL WORLD OF NUMBERS – PYTHAGORAS

The man and the legend

I first heard the name Pythagoras when I was a student in Grade 4 and joined the mathematics club, which was an after-school activity meant for peculiar kids who liked to spend their spare time learning about various strange geometric shapes and developing a rapport with numbers that held within them enigmatic secrets. I liked the way the sound of the name rippled off my tongue: Py-tha-gor-us. It immediately struck me that someone with such a special-sounding name must certainly be a special man. I wasn't wrong. Many consider Pythagoras (c570–c495 BC) to be one of the most fascinating pre-Socratic philosophers. Herodotus (c484–c425 BC), the "Father of History", already pointed to Pythagoras as one of the greatest philosophers of Ancient Greece. Even Heraclitus (c535–c475 BC), a most arrogant philosopher who griped that all humankind was stupid (except for him, of course), admitted that Pythagoras really knew his stuff.

Uncovering solid facts about Pythagoras's life is not easy, mainly because most sources that speak of him were written years after his death. Pythagoras himself, it seems, did not write much – if anything. However, Diogenes Laërtius,[1] the author of *Lives and Opinions of Eminent Philosophers* (biographies of all the greats of Greece), pointed to three books written by Pythagoras: *On Education*, *On the State* and *On Nature.* However, he is pretty alone in this opinion, and most other historians claim that these books were not written by Pythagoras.

By the way, there are many great people who never actually put quill to parchment: Socrates, Buddha, and Jesus, to name but a few. It is just a pity that nobody did for Pythagoras what Plato did for Socrates by writing the dialogues.

There are two biographies of Pythagoras that I prefer above all others. One is by the Neoplatonic philosopher and mathematician Porphyry (c234–c305), and the other is by Diogenes Laërtius. These biographies are frenetically chaotic, amazingly imprecise, and chock-full of contradictions, making them most entertaining and leaving a lot of room for imagination.

There are many legends connected to Pythagoras. Some believed that he was the son of the god Apollo. There are accounts of him having been seen in four different places at the very same time. (Note, however, that Diogenes drops a bombshell at the end of his description of Pythagoras's life by noting that there were, in fact, four men named Pythagoras. That being the case, it does not seem so improbable to observe "Pythagoras" in four different places at the same time.) Some swore that Pythagoras was two-and-a-half metres in height. Others claimed that Pythagoras was an excellent boxer who had never lost a match, and to relax before and after his bouts, he enjoyed playing the lyre while chanting lines from Homer's and Hesiod's poems. Some even declared that Pythagoras's singing cured the ailing. Remarkably, both Porphyry and Diogenes recounted the same story: that one day, as our philosopher strolled along the river, the river greeted him with "Hello, Pythagoras ..."

Now, note Pythagoras's "modesty":

> *There are gods, there are humans, and there is Pythagoras.*
> (Pythagoras)[2]

According to Diogenes Laërtius (assuming we believe him), Pythagoras himself never missed an opportunity to add to the inordinate mystery surrounding his image. For example, he relished telling stories about his "earlier lives", proudly demonstrating how he remembered them all in minute detail. For example, he recalled how he, as the great warrior Euphorbias, participated in the Trojan War. He also spoke about a number of less dramatic previous lives, such as being a successful merchant, a courtier to a king, an animal and even a leaf on a tree. How did Pythagoras know all this? Once he met Hermes, the messenger and emissary of the Greek gods, and as Pythagoras tells it, he was so impressed by our hero that he offered him any gift he might desire, except for the gift of immortality. Pythagoras decided to request the ability to remember all his previous reincarnations. Ask and ye shall receive!

Monotheist philosopher and poet Xenophanes (c570–c475 BC) tells that once, Pythagoras saw someone beating a dog and demanded he cease immediately – because the soul of a friend of his who had recently died resided in the dog. (I hope that man listened to him, since there may still be something to the story that Pythagoras was a great boxer.)

Pythagoras enjoyed playing tricks on people. Once, he disappeared for an extended period of time. When he appeared again in the city, he was just skin and bones. He explained to everyone that he had been away in the land of the dead and recounted not only stories about what he "had observed" among the departed, but also events that the still-living had experienced while he had been absent. How did he do this? The explanation is very simple. He had been hiding in his mother's house while following the diet of a top fashion model. His mother updated him about all the current events, and about what went on in the land of the dead – well, one can say anything of course. Who is going to check the accuracy of the information?

Despite all that is written above, there are still a few facts about Pythagoras about which there is no dispute. And here they are:

Pythagoras was born on the island of Samos in approximately 570 BC. When he was forty years old, Polycrates became "tyrant of Samos" and Pythagoras fled to escape living under his dictatorship. He settled in Croton, in the south of Italy, where he founded the Pythagorean Brotherhood. This had three branches – political, mathematical, and theistic. It has been said that the Brotherhood had close to 300 members at its peak.

Some later biographies of Pythagoras emphasize his great effect on life in Croton. According to the stories, Pythagoras induced the residents of the city to abandon corruption. He transformed them into perfect ascetics who practised brotherly love and sought knowledge, justice, and meaning. Pythagoras was a revered orator, despite the fact that he often spoke behind a curtain in such a way that no one could see his face. (This is a cute trick that certainly adds to his mystery – I'm considering doing the same with my students.)

Before settling in Croton, Pythagoras took several voyages in the pursuit of basic knowledge. He visited Egypt, Phoenicia, Arabia, Judea, and Babylon (then part of the Persian Empire). The Egyptians taught him geometry, the Phoenicians arithmetic, the Chaldeans astronomy, and the Magians, that is Zoroastrians, taught him the principles of religion and practical maxims for the conduct of a good life. He may have even travelled to India (travelling to India was not as popular then as it is now). However, it is not completely clear who the wonderful people he met were and what pearls of wisdom he gathered there.

There are many versions of how Pythagoras died. Most of them are quite dramatic (as you may have guessed). The version I will tell you is the tamest one: Pythagoras died at the age of 90 from natural causes.

About music and numbers

During the time that Pythagoras and his disciples were exploring the laws of the universe, they were also studying the laws of music. You surely recall that Pythagoras loved to chant songs by Homer and Hesiod (*Greatest Hits of Ancient Greece*) while strumming on his lyre. Pythagoras believed that music has enormous influence on the soul and could lead to extremely powerful emotions. (If you doubt this, read *The Kreutzer Sonata* by Leo Tolstoy.) It is clear to us today that the Pythagoreans were greatly affected by Pythagoras's discovery that the musical scales were numerically based. An example of this numeric basis can be observed in many ways. For example, Pythagoras discovered that the lengths of the strings of two notes separated by exactly one octave (e.g. from C to C) have the ratio of 1:2. The strings of two notes a fifth apart (e.g. C–G) have a ratio of 2:3, and the strings of two notes a fourth apart (C–F) have a ratio of 3:4.

> *Music is the pleasure the human mind experiences from counting without being aware that it is counting.*
>
> (Gottfried Leibniz)

Pythagoras's discovery that music can be transformed into mathematical expressions was a major step on his way to his earth-shattering conclusion that the world in its entirety is based, one way or another, on numbers. In fact, Aristotle, in his book *Metaphysics*, points out that the Pythagoreans were the first to study mathematics and conclude that the laws of mathematics governed the laws of all things.

> *Which scientific law guarantees us that there will be scientific laws?*
>
> (Martin Gardner)

Mathematics also governs the visual arts. Perspective is based on geometry and proportion (the sizes of objects represented on a two-dimensional surface decrease in proportion to their distance from the viewer) and the tenets of composition are based on geometrical forms.

> *Geometry is the foundation of all painting.*
>
> (Albrecht Dürer)

But Pythagoras went one step further. He also used the language of geometry to define good and bad, right and wrong. For example, in place of the terms "good" and "bad", he used "straight" and "crooked" (in Greek, of course). We still use the phrase "crooked politician" to refer to someone who takes bribes. A straight line was decent, a crooked one indecent. Echoes of his way of thinking may possibly be found in the expression "an upright individual", since there is certainly no connection between a person's posture and their integrity.

The beginning of a wonderful friendship – amicable numbers

Aristotle once stated that true friends are two bodies sharing one soul. But how did Pythagoras define friendship? Here, we are in for a surprise.

According to neo-Platonic scholar Iamblichus (c245–c325), another author of a Pythagoras biography, the Pythagorean definition of friendship is represented by the numbers 284 and 220.

What?! Why?!

To understand the reason, add together the divisors of 220 (numbers that divide into 220 without a remainder) and then add together the divisors of 284. (Do not include the numbers themselves.)

The divisors of 220 are 1, 2, 4, 5, 10, 11, 20, 22, 44, 55, and 110, and their sum is 284.

The divisors of 284 are 1, 2, 4, 71, and 142, and their sum is (wait for it ...) 220!

According to Pythagoreans, bosom buddies are like a pair of numbers where the sum of the divisors of each is equal to the other number. In mathematics, we call such a number pair "amicable numbers".

A computer can help us determine other pairs of amicable numbers. Alongside (220, 284), there are also (1,184 and 1,210), (2,620 and 2,924), (5,020 and 5,564), and (6,232 and 6,368). These five pairs are the only such pairs below 10,000. If you are really bored, try checking if these pairs really are amicable pairs. In other words, sum up the proper divisors of each (not including the number itself) and see if that sum is the other number in the pair.

If you want, you might like to do something even more challenging – try to find other pairs of amicable numbers. You will probably want to make use of a computer, but keep in mind that in 1636, French mathematics enthusiast Pierre Fermat discovered that (17,296 and 18,416) were a pair of amicable numbers, and that, two years later, renowned French philosopher and mathematician René Descartes discovered the pair (9,363,584 and 9,437,056).

DESCARTES, INFINITY AND GOD

Descartes was deeply impressed by the concept of "infinity", and in his book *Meditations on First Philosophy* he even "proves" the existence of God using the concept of infinity. His proof goes something like this:

Since I myself am a finite creature, I obviously cannot invent the concept of infinity, since only something infinite itself can properly encompass in its thoughts the notion of infinity. Therefore, only God himself can be the creator of the concept of infinity. Since I am able to fathom an infinite God, and since God is the only one who could possibly have been the creator of this notion – it therefore holds that God exists!

In the seventeenth century, there were no computers – never mind the internet and social networks – which make Fermat's and Descartes' accomplishments even more astounding. How did they discover these huge amicable numbers? Read on.

According to some mathematics historians, Descartes did not figure it out on his own. The pair of amicable numbers he presented had actually been determined back in the sixteenth century by Iranian mathematician Muhammad Baqir Yazdi! It is well known that Arab mathematicians were acquainted with quite a number of amicable pairs well before Western mathematicians reconstructed them.

In fact, way back in the ninth century, Iraqi mathematician, astronomer, and physician Thābit Ibn Qurra (826–901) formulated a sufficient condition[3] for a pair of numbers to be amicable.[4] Hundreds of years later, Descartes and Fermat uncovered the formula and used it to make their "discoveries".

It is interesting to note that up till 1866, the second smallest amicable pair, (1,184 and 1,210), had not yet been discovered. This pair was discovered in that year by a young Italian boy named B Nicolò I Paganini (*not* the famous violinist and composer)! It is not clear how every single mathematician since Pythagoras managed to miss this cute little pair. One reason might be that Thābit Ibn Qurra's condition does not apply in this case. Another reason might be that "if you look for nothing, that's what you will find".

By 2007, approximately 12,000,000 pairs of amicable numbers had been discovered. We obviously live in a very amicable world, despite it all.

Feminine numbers, masculine numbers

Alongside his other concepts about numbers, Pythagoras was also convinced that numbers had feminine and masculine traits. For example, odd numbers could be considered feminine and even numbers masculine. Observe now that all the amicable pairs mentioned so far are masculine (even).

This obviously leads to the question, are there also female amicable pairs? Indeed, there are. Here are a few examples: (11,285 and 14,595), (67,095 and 71,145), and (522,405 and 525,915).

This leads us to the ultimate question, is "friendship" between a "male" and a "female" number possible? In other words, is it possible for the sum of the divisors of an odd number and of an even number to equal one another?

Up to the time of writing, no one has discovered an answer. On the one hand, no such pair has yet been discovered; on the other, neither has anyone proven that such a pair is impossible.

At this point, I shall take a break from Pythagoras (I will get back to him later in this chapter), because thinking about "friendly" numbers gets me to thinking about some other anthropomorphic characteristics of numbers, which leads me off on several interesting tangents.

Narcissistic numbers

I have hardly anything in common with myself.

(Franz Kafka)

I am pretty sure that there are some people who can be considered to be in a deep amicable relationship with themselves. But let us think the way Pythagoras might have and instead ask this question with respect to numbers: are there any numbers for which the sum of their proper[5] (positive) divisors equals the number itself?

Numbers that have this property are called "perfect numbers". It becomes immediately (well, after a bit of thinking, of course) apparent that 6 and 28 are the first two perfect numbers. I will pause at this moment so that my wise readers can ascertain that 6 and 28 are, indeed, perfect numbers.

Answer: $1 + 2 + 3 = 6$, $1 + 2 + 4 + 7 + 14 = 28$

Regarding the number 6, here is what Saint Augustine of Hippo (354 –430) wrote in his book *De Civitate Dei* (The City of God): "Six is a perfect number. It is not perfect because God created everything in six days; God created everything in six days because six is a perfect number."

The perfect number that comes after 28 is 496 and the one following is 8,128. Russian author Leo Tolstoy liked to boast that he was born in the "nearly perfect" year of 1828. Had he been born on June 28, he really would have had something to brag about (not to mention that 6.28 is also close to 2π).

You may have noticed a trend here: 6, 28, 496, 8, 128 … People who enjoy coming up with hypotheses might venture the following: the final digit of perfect numbers alternates between 6 and 8.

This hypothesis, however, does not hold. The fifth perfect number is 33,550,336, which does follow the pattern. However, the sixth one – 8,589,869,056 – also ends in six, thus breaking the pattern. Perhaps we have to tweak that hypothesis a bit and suggest instead that all perfect numbers end in either a 6 or an 8.

Let us observe the first *nine* perfect numbers:

6
28
496
8,128
33,550,336
8,589,869,056
137,438,691,328
2,305,843,008,139,952,128
2,658,455,991,569,831,744,654,692,615,953,842,176

That last one has 37 digits. (Yes, the sum of all its proper divisors equals that number itself!)

The tenth number has 54 digits, and the eleventh has 65 and ends with the digits 8,128, which is exactly the fourth perfect number. By the way, perfect numbers with a million (!) digits have been found. Feel free to put forward some hypotheses of your own.

Brain-twister for advanced students

Prove that every *even* perfect number ends with either 6 or 8. If you pay attention to the equations shown below, it may help you in your task.

$$6 = 1 + 2 + 3$$
$$28 = 1 + 2 + 3 + 4 + 5 + 6 + 7 = 1^3 + 3^3$$
$$496 = 1 + 2 + 3 + 4 + ... + 31 = 1^3 + 3^3 + 5^3 + 7^3$$
$$8{,}128 = 1 + 2 + 3 + 4 + ... + 127 = 1^3 + 3^3 + 5^3 + ... + 15^3$$

French mathematician Edouard Lucas (1842–1891) did actually prove that every *even* perfect number must end with 16, 28, 36, 56, 76, or 96. How did he do it? It wasn't easy!

Up to now, we have only seen perfect numbers that are even. So we naturally must ask the following: is there such a thing as an odd perfect number?

At the end of the nineteenth century, British mathematician James Sylvester wrote that the discovery of an odd perfect number would be a miracle. Even today, most mathematicians tend to believe that the answer to this question is "no". Nevertheless, no one has been able to prove this. Here is another "open problem" and another chance for you to acquire fame and glory!

Another interesting question that has not yet been resolved is whether the set of perfect numbers is infinite. Will there always be a higher perfect number no matter how far we progress along the natural numbers? Or is there somewhere a perfect number that is the largest possible?

This is still open and is firmly linked to Mersenne numbers, which we will return to later.

How much does a number weigh? Perfect, fat, and thin numbers

In the spirit of the age of diets, we might say that natural numbers are divided into three categories – perfect, "fat" and "thin". The sum of the proper divisors of "fat" numbers is greater than the number itself, whereas the sum of the proper divisors of "thin" numbers is (I'm sure you can guess …) less than the number itself. For example, 12 is a chubby number, since the sum of its divisors (1, 2, 3, 4, and 6) is 16. The number 10, on the other hand, is skinny, since 1 + 2 + 5 = 8.

What about female numbers? That is to say, odd numbers? Can they also be overweight? Are there odd numbers where the sum of the proper divisors is greater than the number itself? If we experiment a bit, it appears that the sum of the proper divisors of any odd number is always a value *less* than the number itself. (Go ahead and check out some numbers.) If you didn't check any numbers greater than 900, you may have convinced yourself that odd numbers are never fat. However, don't be fooled! Examining a finite quantity of numbers, no matter how many, doesn't mean that the next number won't prove to be the exception. In fact, odd numbers can be fat, as the sum of the proper divisors of 945 is (just add 1, 3, 5, 7, 9, 15, 21, 27, 35, 45, 63, 105, 135, 189, 315) … 975. Thus, we have discovered 945, which is the lowest fat odd number. Nevertheless, overweight odd numbers are pretty rare.

We shall return to the subject of perfect numbers later in the book.

Fascinating and boring people, boring and fascinating numbers

Trying to form "ultimate" lists sometimes ends up creating a type of paradox in which the existence of the definition immediately excludes the thing being defined by that definition from the list. What does this mean?

Let us imagine that we are to prepare two lists. One is a list of the names of *all* the fascinating people in the world, in order of intrigue. The second is a list of the names of everyone else. This, too, will also be in order: from the most boring person in the world to the "simply" uninteresting.

Here are the "top-of-the-lists" of both:

Fascinating people: Pythagoras, Leonardo da Vinci, Cleopatra, Mozart, Einstein, Marilyn Monroe, Socrates, Messalina, Byron, Napoleon, Buddha, Joan of Arc, Alexander the Great ...

Non-fascinating people: Reginald von Yawn, Brunhilda Drowserman, Jacob Soporific, Vladimir Siestakovitch, Bill Boring, Niles Numbstrom, Bernie Blah, Kay Lethargiston, Harry Humdrum, Tim Dimwit ...

However, things are not that simple. Take Reginald von Yawn. According to our list, he is *the most boring* person in the world. This fact, though, changes him into someone interesting. (I mean, what a title to be called the *MOST* boring person in the world!) Therefore, we have to move his name over to the list of fascinating people. Of course, he won't be anywhere near the top of the list, but still, he will have moved into this list and may probably be in quite a respectable place.

Now note what is happening. Since Reginald has left the list, Brunhilda Drowserman has now become *the most boring* person in the world. But this now makes *her* somewhat fascinating, meaning that she also must be moved to the first list. If we continue this process, we will have no alternative but to conclude that there are not, and never were, any boring people in the world. (I'm pretty sure that you long ago discovered the error in this explanation.)

In the world of mathematicians, there is a popular version of the boring people paradox which involves the set of natural numbers that *cannot* be described in *fewer* than 1,000 words. Note that the number of words is finite (exactly 171,476 words appear in the twentieth edition of the *Oxford English Dictionary*) and we have a limited number of words (1,000), and therefore the number of these numbers is finite. Nevertheless, there is such a thing as the smallest natural number that cannot be described in fewer than 1,000 words. We shall denote it by n and define it as: "The smallest number that is indescribable using fewer than one thousand words."

But no! We have just described the number n in 12 words (check it), therefore the number n goes into the list of things that *can* be described in less than 1,000 words, thus negating our definition of the number itself.

The number n and Reginald von Yawn have the same status in the two paradoxes. Both were defined as members of a particular list, but then, by virtue of the definition, they had to be excluded from the list.

What is the sting of these two paradoxes? Mathematics cannot tolerate a paradox and always seeks some explanation to settle the mind. However, in these cases, we must note that we used the non-mathematical attribute of "can be described" without defining exactly what we mean.

This brings us to the next topic of discussion.

Is there actually such a thing as a boring number?

Are there numbers that are more interesting than others and numbers that are boring?

Pythagoras believed that there was no such thing as a boring number, that each and every number is interesting in some way and that each and every number will have either at least one property that makes it unique or some hidden beauty or nuance.

And, indeed, because of the vast importance Pythagoras placed on numbers, he wanted not only to understand them mathematically, but to discover within each some beauty, mystery or secret.

What characteristics of a number make it special or attractive? I guess it is a matter of taste. Is being a perfect number an "attractive" feature? In my opinion, it is. I also believe that a pair of numbers with the amicable property is also interesting – these numbers know what it takes to be a friend. If you wish, you can find beauty in anything, since, as they say, beauty is in the eye of the beholder.

Let's take a look, for example, at 64. The fact that 64 is the square of 8 ($8^2 = 64$) does not really make it special: there are many numbers that are perfect squares. But 64 can also be written as follows: $64 = 2^6 = 4^3$.

Now this is a much more interesting feature. In fact (and it is a fact that is pretty easy to check), 64 is the first number (other than 1) that is not only a square (that is to say, the 2nd power of a number), but is also a 3rd power and also a 6th power.

So, shall we proclaim 64 to be a special number? Might the fact that it represents the number of squares on a chessboard help? There are 64 positions in the *Kama Sutra*, and 64 hexagrams in the *I Ching*, the Chinese *Book of Changes*. Do these facts make 64 any more special? I'm not sure – you decide. And feel free to see if you can find any other things that make 64 unique.

If we assume that every number has some interesting feature, let us examine 65, which is the very next number after 64. Can we find anything remarkable about it?

We certainly can! This number is the second number in the set of natural numbers (after 50) that can be presented in two different ways as the sum of two squares: $65 = 8^2 + 1^2 = 7^2 + 4^2$. *And*, it can also be written as the sum of two cubes! $65 = 1^3 + 4^3$. In fact, 65 is the first number that can be written both as the sum of two squares (two different ways!) and the sum of two cubes. Astounding!

Pythagoras himself thought that one of the most interesting numbers was 36. First of all, he was convinced that this was the ideal age for a man. (I don't know if he ever gave an opinion about what the ideal age for a woman was.)

The number 36 impressed Pythagoras mathematically because:

$$36 = (1 + 2 + 3)^2 = 1^3 + 2^3 + 3^3$$

When I was younger, I concurred with Pythagoras (both about the ways that 36 could be constructed and *also* that it was a pretty nice age to be at), but today I take a more optimistic approach and my new "ideal age" – for both men and women – is 100: $100 = (1 + 2 + 3 + 4)^2 = 1^3 + 2^3 + 3^3 + 4^3$

The equations we just discussed are not by chance. As you may have already guessed, the sum of any number of consecutive numbers squared is equal to the sum of the individual numbers cubed:

$$(1 + 2 + ... + n)^2 = \left[\frac{n(n+1)}{2}\right]^2 = 1^3 + 2^3 + 3^3 + ... + n^3$$

We have discovered some interesting properties about some numbers. But certainly there must be some numbers that don't have anything really unique about them. However, if we apply the "most boring person in the world" paradox to numbers, perhaps a number without any special properties can still be considered "interesting" simply because of this feature.

LESSON 2

RAMANUJAN AND PYTHAGORAS'S PEBBLES

Part 1: A man who knew infinity

INTERMEZZO

The passage to India: When Hardy met Ramanujan

Srinivasa Ramanujan was a mathematical genius. Born in 1887 in the Erode district in Madras, India, he demonstrated unusual mathematical prowess at a young age.

However, there was no one with whom he could study nor even to advise him *what* to study where he lived. One might say that Ramanujan was an autodidact. Without any formal training, he reached unprecedented achievements in several mathematical disciplines.

The main focus of his work was number theory, and, like Pythagoras, Ramanujan also had an intimate relationship with numbers.

In 1913, Ramanujan sent several of his mathematical results (equations) to three notable British mathematicians, but only one of them, Godfrey Harold Hardy, grasped the brilliance of the person behind the results before him. Although the results were very much a diamond in the rough, they were still of rare beauty. Hardy made a point of bringing Ramanujan to London and then to Cambridge University during the First World War. Ramanujan later became the first Indian elected as a fellow of Trinity College, Cambridge.

Below are two of the results (equations) that amazed Hardy so much. When I first saw these equations, I was a third-year mathematics undergraduate, and they were so beautiful that I immediately thought of music. They looked to me like the notes of a beautiful symphony. The equations seem really complicated, and they are, but you don't need to understand them. You don't even have to look at them as mathematics. Just see the glorious aesthetic beauty embodied in the patterns of the numbers.

RAMANUJAN'S 1ST SYMPHONY

$$1-5\left(\frac{1}{2}\right)^3+9\left(\frac{1\times 3}{2\times 4}\right)^3-13\left(\frac{1\times 3\times 5}{2\times 4\times 6}\right)^3+\cdots=\frac{2}{\pi}$$

$$1+9\left(\frac{1}{4}\right)^4+17\left(\frac{1\times 5}{4\times 8}\right)^4+25\left(\frac{1\times 5\times 9}{4\times 8\times 12}\right)^4+\cdots=\frac{2\sqrt{2}}{\sqrt{\pi}\,\Gamma^2\left(\frac{3}{4}\right)}.$$

$$\int_0^\infty \frac{1+\frac{x^2}{(b+1)^2}}{1+\frac{x^2}{a^2}}\times\frac{1+\frac{x^2}{(b+2)^2}}{1+\frac{x^2}{(a+1)^2}}\times\cdots dx=\frac{\sqrt{\pi}}{2}\times\frac{\Gamma\left(a+\frac{1}{2}\right)\Gamma(b+1)\Gamma(b-a+1)}{\Gamma(a)\Gamma\left(b+\frac{1}{2}\right)\Gamma\left(b-a+\frac{1}{2}\right)}.$$

$$\int_0^\infty \frac{dx}{\left(1+x^2\right)\left(1+r^2x^2\right)\left(1+r^4x^2\right)\ldots}=\frac{\pi}{2\left(1+r+r^2+r^6+r^{10}+\cdots\right)}.$$

How exquisite!

An equation for me has no meaning, unless it expresses a thought of God.

(Ramanujan)

While it is possible to simply enjoy the aesthetic appearance of Ramanujan's mathematical equations, perhaps we wish to be pedantic and find out if the results are indeed correct.

Let us look at the first equation.

We have here an infinite series of addends separated by alternating pluses and minuses. The first number is 1, but then each subsequent addend is a product of an integer number and a fraction. The integer number increases by 4 each time. In the fraction, the numerator is the product of odd numbers and the denominator the product of even numbers, and the quantity of factors in each increases by one each time. Ramanujan claimed that the more addends in the equation, the closer it gets to 2 divided by pi (the ratio between the circumference of a circle and its diameter)! The result will be exactly 2 divided by pi for an infinite number of addends.

Where did this equation come from? Ramanujan had thousands (!) of such equations (almost 3,900 to be more precise). You probably won't believe this, but the ones above are some of his simplest!

To be perfectly honest, I must disclose that a few of Ramanujan's formulas were not 100 per cent correct, but I stand firm in the belief that one can learn much more from the mistakes of great people than from the exactitudes of mediocre ones.

Hardy and Ramanujan

Hardy and Ramanujan were extremely different from one another in character. Hardy was an atheist (he saw God as his greatest enemy) and extremely pedantic in matters of mathematics – he wanted to see a proof for every equation. Ramanujan, on the other hand, was deeply religious in every sense and relied more on intuition when it came to mathematics. Not only did he see God's expression in his equations and identities, he actually hesitated to reveal how they came to him for fear that he would be committed on grounds of insanity. This reminds me of the scene in Miloš Forman's movie *Amadeus* where Salieri reads the score of Mozart's *Gran Partita* and is certain that God himself dictated the notes to Mozart. Salieri prattles on about how God didn't see fit to dictate such a lofty composition to him. I guess that there are some who believe that genius must come from God.

I think that the following equation is the strangest Ramanujan formula:

$$1 + 2 + 3 + 4 + \cdots = -1/12$$

Really??? It seems so incorrect! The infinite sum on the left must add to infinity and by no means can it be a negative number! But make no mistake – Ramanujan knew what he was doing and had a reason for writing it down: he was dealing with the highly important Riemann–Euler zeta function (this is a function of a complex variable and beyond the scope of this book). In a letter to Hardy, Ramanujan wrote: "The sum of an infinite number of terms of the series $1 + 2 + 3 + 4 + \cdots = -1/12$ under my theory. If I tell you this you will at once point out to me the lunatic asylum as my goal."

Despite Hardy's rigidity concerning mathematics (in other matters he was known as extremely kind-hearted), he could not help but be captured by the charming equations of this Indian genius.

Ramanujan's formulas must be true because, if they were not true, no one would have the imagination to invent them.

(G H Hardy)

Hardy showed Ramanujan's work to one of his colleagues with whom he often worked, John Littlewood (we met him earlier with the paradox of the tennis balls). Littlewood was also astounded by Ramanujan's apparent genius. He commented that he was not aware of any mathematician with whom he could compare Ramanujan – he surpassed them all.

> *To illustrate to what extent Hardy and Littlewood in the course of the years came to be considered as the leaders of recent English mathematical research, I may report what an excellent colleague once jokingly said: "Nowadays, there are only three really great English mathematicians: Hardy, Littlewood, and Hardy-Littlewood."*
>
> (Harald Bohr)[1]

Hardy was an extremely noteworthy mathematician. But when Paul Erdős (we also met him earlier) asked Hardy what he believed was his greatest contribution to mathematics, Hardy answered, "Discovering Ramanujan."

I will just add this: Hardy was in the habit of rating mathematicians on a scale from 0 to 100. He gave himself a rating of 25, gave his colleague Littlewood the rating of 30, and rated the great German mathematician David Hilbert (after whom the branch of mathematics known as "Hilbert Spaces" is named) an 85. To Ramanujan, he gave the ultimate rating of 100!

More on Hardy and mathematical thinking

One of my favourite books is Hardy's *A Mathematician's Apology*. It discusses the aesthetics of mathematics and allows a rare glimpse into the thinking patterns of those involved in this discipline. Hardy loved pure (theoretical) mathematics and once even boasted that nothing he did had any practical purpose. Here, though, he greatly erred. For example, anybody who has ever studied anything about population genetics is familiar with the Hardy–Weinberg principle. Hardy also thought that number theory – which he loved dearly – had no useful purpose. Today, number theory is intimately connected to ciphers and codes. Hardy even thought that the theory of relativity had no practical use. It is indeed very difficult – perhaps even impossible – to predict which mathematical discoveries will prove useful in practice and which ones will be useful "only" for nurturing the glory of the human mind.

In his book, Hardy explains in the most fascinating way what he considers beautiful in mathematics and what not. We will discuss this later on.

THE RAMANUJAN PRIZE

Unlike matters of mathematics, Ramanujan did not excel in matters of health. He died in 1920, at the very young age of 32, not long after he returned to India.

Beginning in 2005, a prize named after him, the SASTRA Ramanujan Prize, has been awarded annually for discoveries made as a result of his work. The prize is awarded to mathematicians who are no older than 32 – the age at which Ramanujan ended his romance not only with life, but with the numbers he loved so much.

In 2009 (the year the first edition of this book was written), the German mathematician Kathrin Bringmann won the prize. The latest prize at the time of writing, in 2017, was awarded to Maryna Viazovska, a Ukrainian mathematician who solved problems in dimensions 8 and 24!

It is time to return to interesting numbers, which were discussed in the previous lesson.

Taxicab 1,729

One day, Hardy came to visit an ailing Ramanujan. Hardy mentioned that he had come in a taxi whose number was 1,729. "What an extraordinarily boring number," quipped Hardy. "Not at all!" countered Ramanujan passionately. "1,729 is really a *most* interesting number! Do you not realize that it is the smallest number that can be written as the sum of the cubes of two positive integers *two* different ways? The first way is 1 cubed plus 12 cubed. The second way is the sum of 10 cubed plus 9 cubed." Here, let me write it down for you:

$$1{,}729 = 12^3 + 1^3 = 10^3 + 9^3$$

When I recount this story to my friends, they are generally astounded by the fact that someone can immediately calculate that it is possible to write 1,729 as the sum of two cubic numbers. Frankly, I am more astounded by the fact that Ramanujan knew that 1,729 was the *smallest* number with this property. How could he possibly have known this? I have no idea!

(We are, of course, speaking here only about positive numbers. If we could use negative numbers, we can find a value smaller than 1,729. For example, $91 = 6^3 + (-5)^3 = 4^3 + 3^3$.)

> *Each of the positive integers was one of Ramanujan's personal friends.*
>
> (John Littlewood)

I would like to point out that 1,729 has several other interesting properties. This one, my favourite, comes from Japanese mathematician and author Masahiko Fujiwara (born 1943),[2] who showed that 1,729 is one of only three numbers with the following property: the sum of its digits multiplied with the reverse of that sum will produce the original number.

$$1 + 7 + 2 + 9 = 19$$
$$19 \times 91 = 1{,}729.$$

Brain-twister

Find the other two numbers. (The number 1 also has this property, but it is too trivial to count.) Hint: One of the numbers is a double-digit number and it is not too hard to figure out. The second number has four digits.[3]

Kaprekar discovers the secret of 6,174

Indian mathematician Dattathreya Ramchandra Kaprekar was born in 1905. He graduated from the University of Mumbai and devoted himself to a teaching career. He was a schoolteacher for decades, but never went on to study higher mathematics. He made contributions in various topics such as magic squares, recurring decimals, and integers with special properties. He discovered some remarkable properties about numbers yet was never recognized in his lifetime. Only in recent years has his contribution to number theory become appreciated – belated recognition has come in the form of a constant named after him.

Kaprekar's constant

In 1949, Kaprekar discovered that 6,174 can be considered the limit for the sequence of the following actions: take any four-digit number where the numbers are not *all* identical. Rearrange those digits to form the smallest and largest numbers possible. Deduct the smaller from the larger. If the result is 6,174, you are done. If not, repeat the process. 6,174 will be waiting at the end.

Let us try this with the year I originally began this book, 2009. The highest number that can be formed from the digits is 9,200 and the smallest is 0029. Subtract 29 from 9,200 to get 9,171.

We repeat the process: 9,711 – 1,179 = 8,532.

We continue: 8,532 – 2,358 = 6,174. We have arrived at the end of our quest: 6,174 was waiting for us there all along.

In mathematical language, 6,174 is called a "fixed point", which means that if we subject this number to the process, we will return to 6,174. Let us check it: 7,641 – 1,467 = 6,174. There is, indeed, nowhere else to go; the journey has reached its end.

What if we decide to get a little clever? Would this work with a number with three identical digits? Say 1,112? Let's give it a go.

$$2{,}111 - 1{,}112 = 999$$

(Since we need to work with four-digit numbers, we will write the result as 0999.)

$$9{,}990 - 0999 = 8{,}991$$
$$9{,}981 - 1{,}899 = 8{,}082$$
$$8{,}820 - 0288 = 8{,}532$$
$$8{,}532 - 2{,}358 = 6{,}174$$

We have arrived.

If any of you are desperately in need of some occupational therapy, go ahead and try out some other numbers.

We now have an excellent opportunity to conduct a little mathematical experiment of our own. What would happen if, instead of four-digit numbers, we used three-digit numbers?

Let's try, using 169 as our example.

961 – 169 = 792 (By the way, 169 = 13^2 and 961 = 31^2. But I digress.)
972 – 279 = 693
963 – 369 = 594
954 – 459 = 495

We have arrived at a fixed point (check it!). Have we discovered the Kaprekar constant for three-digit numbers? We have indeed! If you are an algebra enthusiast, you can prove this without much of a heroic effort.

Let us move on to two-digit numbers. This should be easy, no?

Let's start with one of my favourite numbers: 17.

71 – 17 = 54, 54 – 45 = 9, 90 – 9 = 81, 81 – 18 = 63, 63 – 36 = 27, 72 – 27 = 45, 54 – 45 = … Hey! We have already been here! What's going on? In fact, we have reached the point of periodicity. With two-digit numbers, there is no fixed point.

Brain-twister

What happens with five-digit numbers? Six-digit numbers?

Kaprekar numbers

Kaprekar discovered that some numbers have an unusual property: if we square the number, the resultant number can be split into two numbers whose sum will be the original number. This concept will be clearer with some examples:

$9^2 = 81$	$1 + 8 = 9$
$45^2 = 2{,}025$	$20 + 25 = 45$
$999^2 = 998{,}001$	$998 + 001 = 999$
$2{,}728^2 = 7{,}441{,}984$	$744 + 1{,}984 = 2{,}728$
$818{,}181^2 = 669{,}420{,}148{,}761$	$669{,}420 + 148{,}761 = 818{,}181$

The numbers 9, 45, 999, 818,181 – and there are many others – are members of the Kaprekar Number Association. A simple program that you can run on your computer can introduce you to many more members.

Brain-twister

Prove that the numbers 9, 99, 999 and 9,999 ... are Kaprekar numbers.

An ancient Indian riddle

Find the next number in this series: 1, 2, 4, 8, 16, 23, 28, 38, 49...

Take a few minutes to think. If you can't figure it out, the answer is in the endnotes.[4]

An interesting fact about this riddle is that, in general, respectable mathematicians have a hard time solving it because they are looking for some complicated idea. Clever children are the ones who handle this riddle best.

Kaprekar observed that some numbers can be obtained by using a smaller number added to the sum of its digits, whereas for others this was impossible. For example, we can get 40 using the method by using 29 ($2 + 9 = 11$, $29 + 11 = 40$). But we can't get to 20 using this method starting with any number. (Check it out.)

Kaprekar formulated a criterion for identifying which numbers *cannot* be obtained through this method. I wouldn't want to deprive you of the pleasure of recreating this criterion. To help you out, find the first number that satisfies this criterion and then see if you can come up with the rule.

And now, we return to our great hero, Pythagoras.

Part 2: Pythagoras on the beach

Imagine that, instead of learning in a school, you had your lessons on the beach. Awesome, right? This is exactly what the Pythagoreans did. Pythagoras loved to use marbles or pebbles placed on the sand to represent numbers.

By arranging the pebbles in different ways, he came up with a number of mathematical formulas and concepts.

We shall look at some examples.

The sum of consecutive odd numbers

Anyone who remembers anything from their school days will probably recall the following law: the sum of the first n consecutive odd numbers, beginning with 1, will always give the square of n.

To illustrate:

$$1 + 3 = 4 = 2^2$$
$$1 + 3 + 5 = 9 = 3^2$$
$$1 + 3 + 5 + 7 = 16 = 4^2$$

and so forth.

Those who made the effort to study mathematics at a high level in high school probably know that this law can be proven using a concept known as "mathematical induction".

Mathematical induction is a most amazing tool for proving concepts and, remarkably, allows one to arrive at a proof for infinite elements by way of a proof for a finite number. I shall elaborate to illustrate how induction works. Assume that we want to prove the following claim for all the natural numbers:

$$1 + 3 + \ldots + (2n - 3) + (2n - 1) = n^2$$

The proof is divided into two parts. For the first part, we prove what is known as the *induction step*, meaning that we prove the somewhat peculiar statement that goes: "If this equality is true for n, then the equality is also true for $n + 1$."

In the second part, called the *base case*, we verify that the equality holds for $n = 1$.

And that's it! We have proved the claim for all the natural numbers.

This might look doubtful, but let me explain. Think of the proof for n as a domino. Anyone who has ever stacked a row of dominoes knows that the dominoes are arranged in such a way that if a particular tile falls, it will knock over the next tile, which in turn knocks over the one after that, and so on, so that eventually all the tiles fall over. Similarly, in induction, we place all of our "arguments" in a row in such a way that if we prove the argument for any element, n, then it "knocks over" the argument for $n + 1$. But just as with the dominoes, to start the chain of falls, we must knock down the first tile, which is simply the base case in induction terms. We perform the induction step: that is, we accept the correctness of this equation:

$$1 + 3 + \ldots + (2n - 3) + (2n - 1) = n^2$$

We now prove the equation for $n + 1$ as follows.

The left side looks like this:

$$1 + 3 + \cdots + (2(n + 1) - 3) + (2(n + 1) - 1)$$
$$= 1 + 2 + \ldots + (2n - 1) + (2n + 1)$$

And the right side is $(n + 1)^2$. Because we determined the correctness of the equation for n, we can now argue that $1 + 3 + \cdots + (2n - 1) + (2n + 1) = n^2 + (2n + 1) = (n + 1)^2$

This ends the induction hypothesis. All that remains is to knock down the first tile. For the base case $n = 1$ the argument is certainly correct since $1 = 1^2$.

Now, one after the other, the tiles of proof fall down: the claim for $n = 2$ is derived from the claim for $n = 1$, and the claim for $n = 3$ from that for $n = 2$, and so on.

Pythagoras, however, had a better way. By simply arranging pebbles in a certain way, the law becomes instantly apparent.

One marble and three marbles are easily arranged into a 2×2 square:

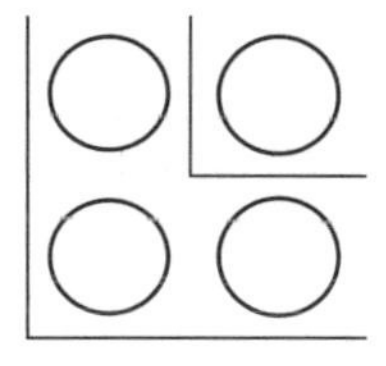

One marble, three marbles, and another five marbles yields a perfect 3×3 square:

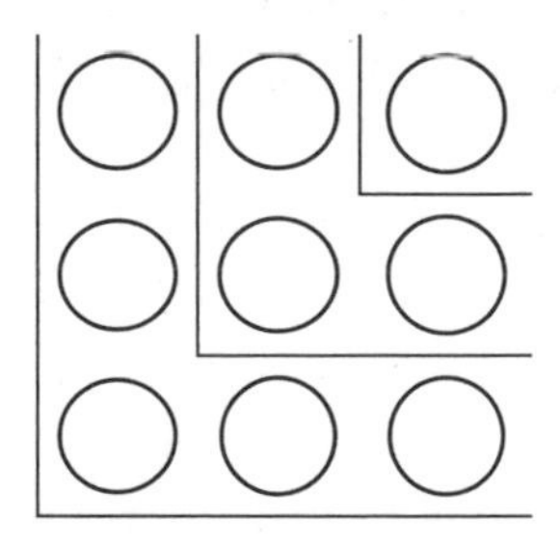

Similarly, adding the next odd number, 7, produces a 4 × 4 square:

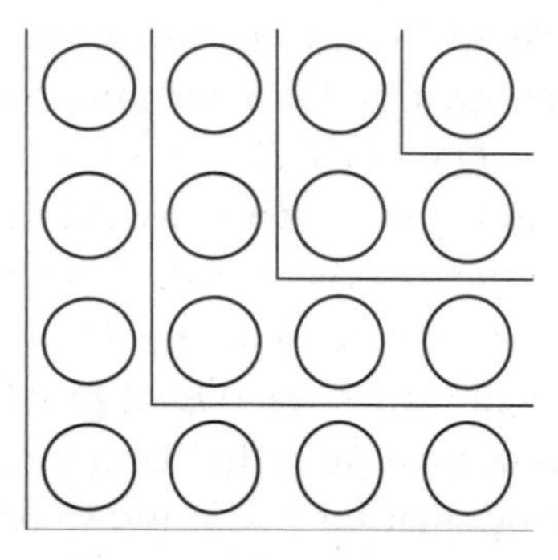

The great Jewish philosopher Baruch Spinoza differentiated between three types of knowledge:

1 Belief
2 Examination (experimentation)
3 Understanding

I'll explain. If you tell me something, such as that the sum of a sequence of odd numbers will yield a square, I may believe that you know what you're talking about. This is the first level of knowledge. However, it is quite possible that what you have told me is incorrect.

If I bother to check it out – that is, to examine a few examples and prove to myself that they work – I now have reached the second level of knowledge. It is a bit more trustworthy, because I have seen that it does work for some cases, but I cannot trust it absolutely. Professor Beno Arbel (1939–2013) once demonstrated to me an impressive example of how repeated testing, even done an extraordinary number of times, does not attest to the correctness of something. Look at the expression $991n^2 + 1$. Is there any value of n for which this expression yields a perfect square? We can check umpteen values for n, and then more values for n, and it will appear that this expression will never yield a perfect square. But this is not true, because if $n = 12{,}055{,}735{,}790{,}331{,}359{,}447{,}442{,}238{,}767$, the answer *will* be a perfect square! We could live for a billion years and spend the entire time testing, and it is doubtful that we would discover this number.

Which brings us to the third level: it is only by understanding *why* something happens – as in the case of arranging stones in the shape of a square – that the chance of erring becomes zero.

> *Tell me and I forget. Teach me and I remember. Involve me and I learn.*
>
> (Benjamin Franklin)

I love Pythagoras's approach because it offers knowledge of the third sort. I understand on a deep level the reason for the correctness of the expressions. I can't check the whole infinity of cases for a formula, but if I have a deep understanding of what is happening, I will understand why the formula is true.

Once, in a library, I happened on the book *Theory of Equations* by Russian mathematician Victor Uspensky (1883–1947). He studied at Stanford University under the name James Uspensky. Uspensky proved all sorts of formulas the way Pythagoras did – that is to say, through the use of illustrations.

I will begin with a fairly simple example.

If we add together all the numbers from 1 to n, the result will be

$$\frac{n(n+1)}{2}.$$

The diagram below explains why this formula works for $n = 4$.

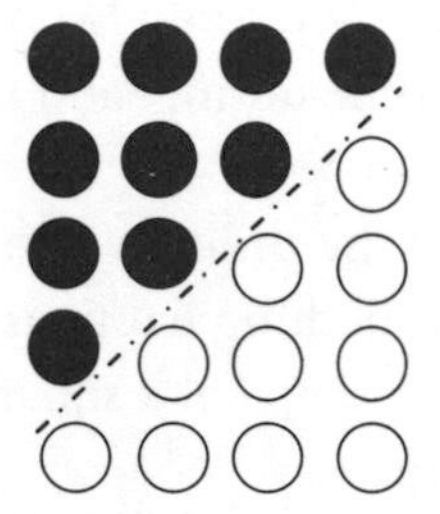

The sum of the numbers from 1 to 4 equals half the area of the rectangle, in other words, $1/2 \times 4 \times 5 = 10$.

Well, that's easy to do with $n = 4$. But what happens with larger numbers?

There is a clever way to calculate the sum of consecutive numbers from 1 to, say, 100. This method is closely associated with the story in which the main hero is a little boy. Different countries and nationalities quarrel over the exact identity of the boy. The Russians claim that it was mathematician Nikolai Lobachevsky, the "Copernicus of Geometry", when he was seven-and-a-half years old. The Jews like to say that it was Baruch Spinoza, also at that age. The Germans place the eminent mathematician – really, one of the greats in the history of mathematics – C F Gauss (the Gaussian bell distribution is, not surprisingly, named after him) at centre stage at the age of six. And there are not a few parents who claim that this happened to their very own child.

I'll go with the Spinoza version since we have just made his acquaintance in this book.

One day, little Baruch was sitting in class and he was very, very bored. However, the problem was that not only was he bored, he was acting up and disrupting the class. The teacher decided to give the boy something to do that would occupy him for a long time, so the teacher asked Baruch to add up all the numbers from 1 to 100. "That should keep him busy at least until the end of the lesson," the teacher said to himself.

Well, expectations are one thing and reality another. The teacher had barely turned back to the blackboard when Baruch called out, "Teacher, the answer is 5,050."

We can assume that Baruch was not aware of the formula above (he was too young), so how could he possibly have arrived at the sum so quickly?

$$1 + 2 + 3 + 4 + \ldots\ldots\ldots\ldots + 98 + 99 + 100 = ?$$

The answer is really quite simple and also quite elegant. Instead of adding all the numbers in order, Baruch realized that adding the first number to the last ($1 + 100 = 101$), and the second to the second-last ($2 + 99 = 101$), and the third to the third-last ($3 + 98 = 101$), and so on, all the way up to $50 + 51 = 101$, gave fifty pairs, each of which came to the sum of 101. So all he had to do was calculate 50 times 101, which is really very easy: $50 \times 100 = 5{,}000$, plus one times 50, for a grand total of 5,050.

Clever, isn't it? If we think about it for a few seconds, we can understand that little Baruch's method is analogous to Pythagoras's idea of arranging the pebbles.

Pythagoras's custom of teaching using pebbles also explains the terms "square", "triangular", "cube" and more. They are simply because he gave those numbers names that matched their geometrical arrangements.

For example, as you can see in the illustration, 1, 4, 9, 16, 25 … are "square" numbers:

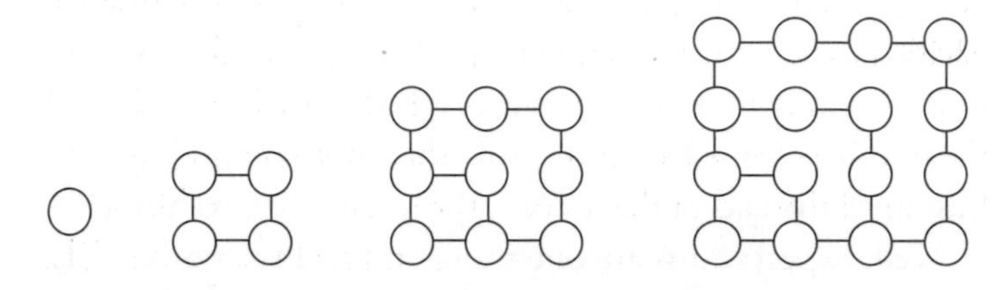

The numbers 1, 3, 6, 10, 15 … are "triangular" numbers:

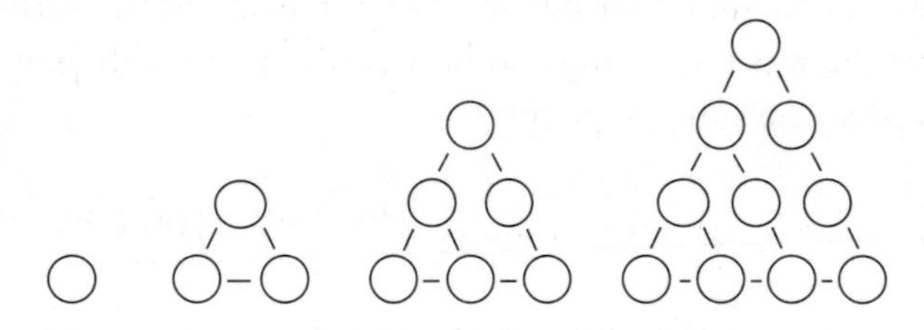

And 1, 5, 12 … are "pentagonal" numbers:

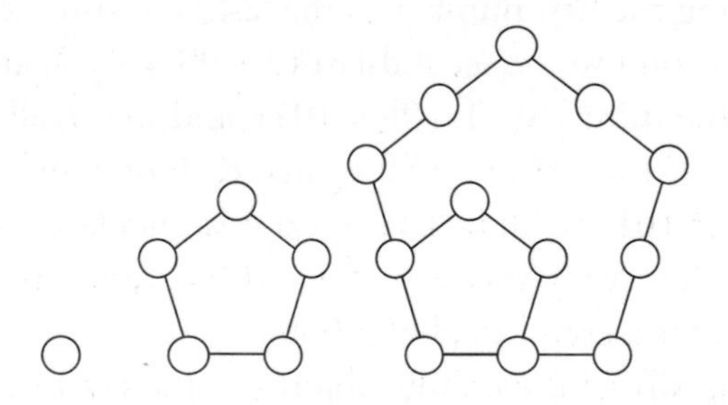

Let's return to triangular numbers.

Theory: Any triangular number from 3 and up can be written as the sum of a square plus two more triangular numbers.

Proof: An illustration should do.

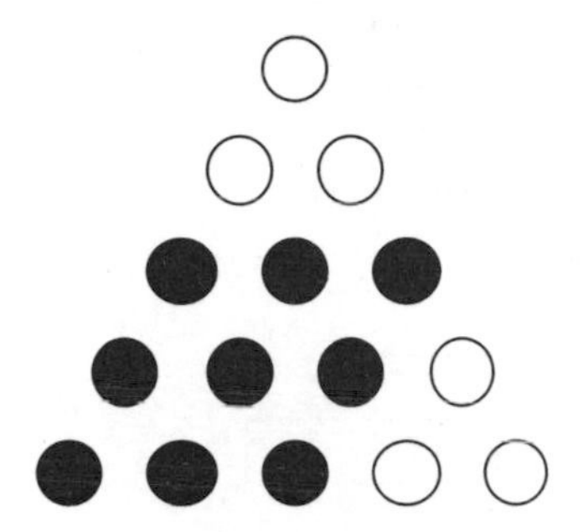

And, indeed, $15 = 9 + 3 + 3 = 3^2 + 3 + 3$

QED.

A tiny note: This theorem, of course, can be proved in the conventional way, but it is a little bit more difficult and a little bit harder to follow. The illustration is so much clearer.

What about the sum of consecutive squares?

$$1^2 + 2^2 + 3^2 + 4^2 + 5^2 + \ldots = ?$$

Someone who really liked induction exercises in school may even remember this formula (with • being the multiplication sign):

$$1^2 + 2^2 + 3^2 + \ldots.. + n^2 = \frac{n \bullet (n+1) \bullet (2n+1)}{6}$$

This equation was already known by Chinese mathematician Yang Hui, who lived in the thirteenth century.

What is the logic behind it? To comprehend it, we have to turn to Pythagoras for a little help. I will demonstrate the concept behind $n = 4$. The general concept follows the exact same principle.

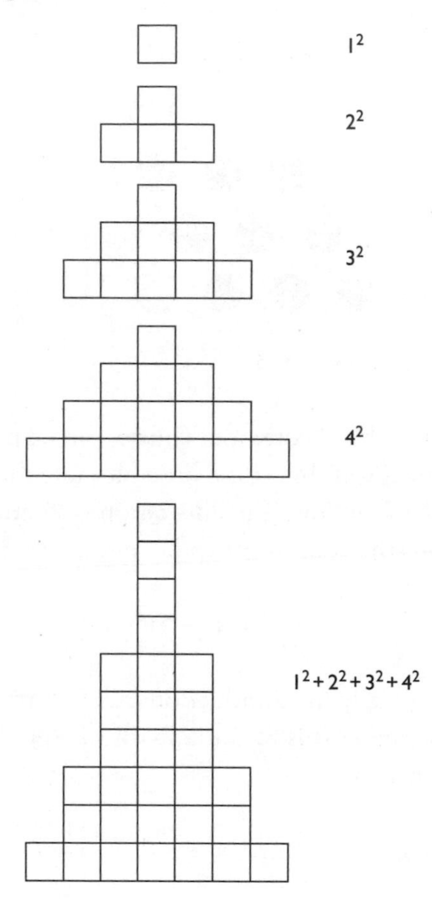

All that is now left to do is to construct the following shape:

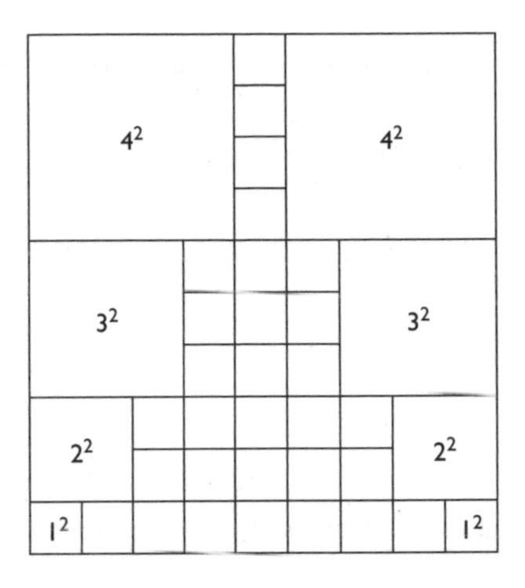

and we have

$$1^2 + 2^2 + 3^2 + 4^2 = \frac{4 \cdot 5 \cdot (2 \cdot 4 + 1)}{6}$$

or

$$1^2 + 2^2 + 3^2 + \ldots\ldots + n^2 = \frac{n \cdot (n+1) \cdot (2n+1)}{6}$$

If you enjoy this "pebbles on the beach" method of mathematics, an interesting diversion might be to find similar formulas in your textbooks, gather together an impressive pile of pebbles, and head to the beach. (Don't forget a good supply of sunscreen – these exercises may take some time.) On the other hand, another entertaining diversion might be to take your pile of pebbles and invent your own formulas. And there is, of course, the ultimate diversion – to go to the beach and not do a thing!

Without mathematics we cannot penetrate deeply into philosophy.
Without philosophy we cannot penetrate deeply into mathematics.
Without both we cannot penetrate deeply into anything.
(Leibniz)

Pythagoras's theorem

Can one write about Pythagoras without mentioning the renowned theorem that bears his name? Of course not. Therefore, I shall end this chapter with some anecdotes about it.

So, as the theorem goes, for any right-angled triangle, the square of the hypotenuse is equal to the sum of the squares of the other two sides.

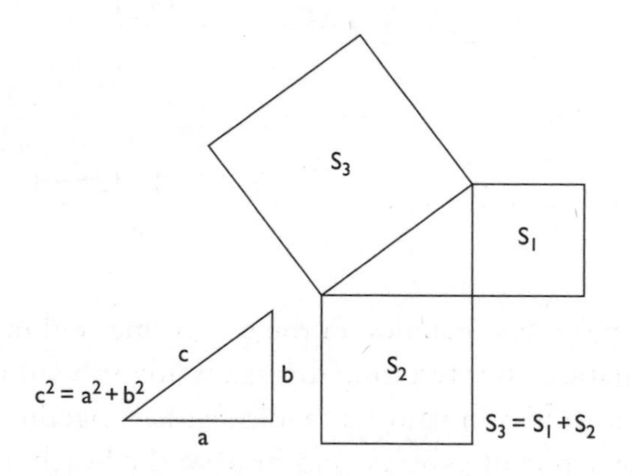

An interesting fact: the ancient Egyptians were aware of this theorem and even used the 3-4-5 Pythagoras triangle to construct right angles.

Surely anyone who has attended school knows how to recite the famous theorem, but how many of you actually know why it works and how to prove it?

The story is told that when philosopher Thomas Hobbes (1588–1679) happened to see Pythagoras's theorem in a book by father-of-geometry Euclid, he was so astounded that he refused to believe that it could possibly be correct. At that time, Hobbes was about 40, and up till that point in time had not been particularly interested in mathematics. Hobbes read the proof (pretty impressive for someone who did not live and breathe shapes) and fell in love with geometry.

Well, if Hobbes refused to believe the veracity of the theorem, I have no alternative but to prove it. Actually, there are hundreds of proofs, beginning with the very first one that Euclid wrote and that so amazed Hobbes, and culminating with proofs that use differentials.

I'll show you three proofs that I particularly like, but before I do, I will demonstrate a proof for the case when the right triangle is isosceles. The proof is so simple that a diagram will suffice.

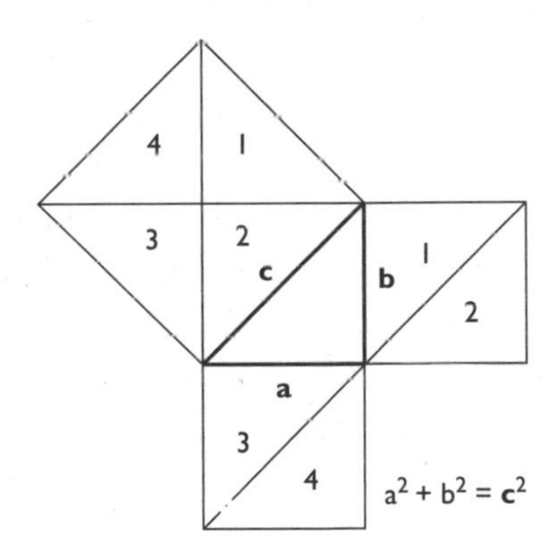

Proof 1 – Beauty in simplicity

I chose this one because it is one of the simplest.

We take a square with side length $a + b$, and arrange in

it four identical right-angled triangles as in the diagram below on the left. Now, we rearrange the four triangles differently, as in the diagram on the right. The shaded areas must be equal in both squares, because they are both equal to the total area of the square minus the area of the four triangles.

Therefore, $a^2 + b^2 = c^2$.

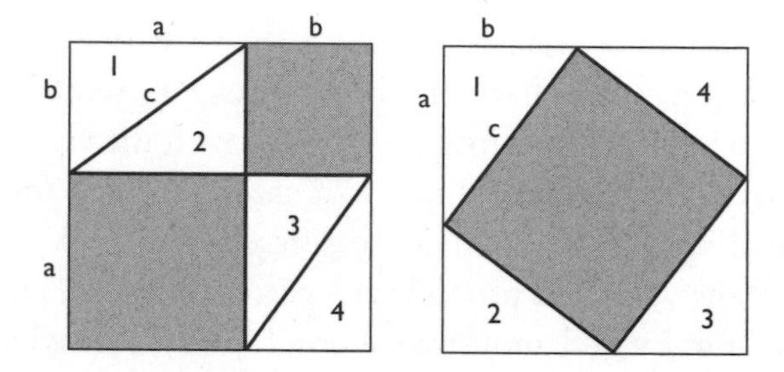

Proof 2 – Garfield's proof

If you are thinking that Garfield the lazy cat is responsible for developing a proof for Pythagoras's theorem, you are wrong. This proof belongs to the twentieth president of the United States, James A Garfield. (However, if you also think that there is no connection between Garfield the cat and President Garfield, you are wrong again! Garfield the cat was named so by his creator Jim Davis, who named him in honour of his grandfather, and his grandfather was named in honour of President Garfield.)

Here is the proof. Observe the following diagram of a trapezoid:

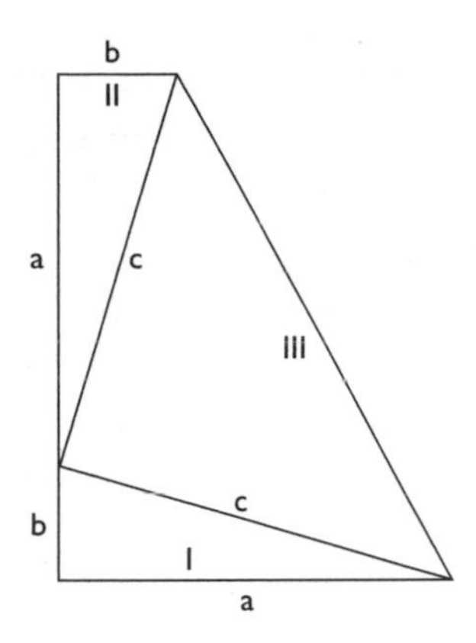

The area of this trapezoid is equal to its height (a + b) times the average of its bases: $\frac{a+b}{2}$. On the other hand, the trapezoid is composed of three triangles – I, II and III – the area of I and II is $\frac{a \times b}{2}$. Now a simple angle-chasing argument can be used to prove that triangle III is right-angled, remembering that the sum of the angles in a triangle is 180 degrees. It follows that the area of triangle III is $\frac{c^2}{2}$, Thus:

$$(a + b) \bullet \frac{a+b}{2} = \frac{a \times b}{2} + \frac{a \times b}{2} + \frac{c^2}{2}.$$

By expanding the left side and simplifying, we end up with $a^2 + b^2 = c^2$.

QED.

Bravo to the 20th president of the United States!

We now await with bated breath for a proof of Pythagoras's theorem by the 45th president of the United States, who is serving at the time I am writing these words, Donald Trump. Tweet it, Donald!

Proof 3 – Pythagoras meets Leonardo

This time, the one responsible for the proof is none other than Leonardo da Vinci. Giorgio Vasari (1511–1574), in his book about the great artists (*Lives of the Most Excellent Painters, Sculptors, and Architects*), tells that Leonardo studied mathematics for only a few months, but it was enough time to make him an expert. Leonardo also didn't invest a lot of time in the study of music, yet, similar to Pythagoras, he enjoyed singing while accompanying himself on the lute. There are many other subjects in which Leonardo did not invest a lot of time and still managed to become expert enough to apply them, even more so than many who did invest a lot of time and effort in studying things.

I am sure that no one will be surprised if I say that the main concept behind da Vinci's proof can easily be understood by a diagram.

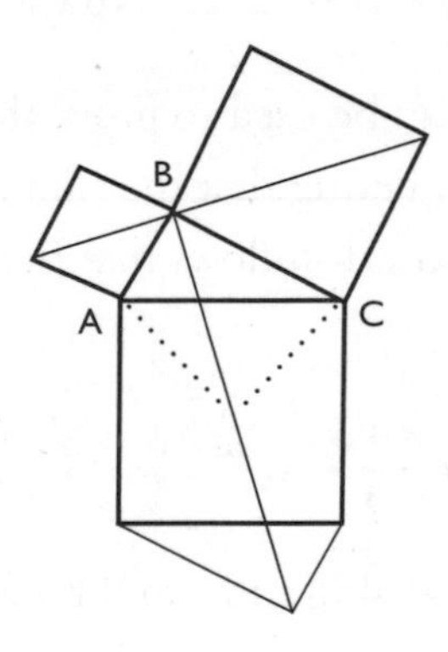

How did Leonardo da Vinci come up with this diagram? Where is the proof hiding? I will give you the time to exercise your brain a bit.

LESSON 3
THE SECRET LIVES OF PRIME NUMBERS

Euclid's aha moment

In the previous chapter, where we met Pythagoras, we were introduced to triangular numbers, square numbers, and even pentagonal numbers. However, we didn't talk at all about rectangular numbers. Why not?

The reason is that they are not very interesting due to their abundance on the market. Any number that is divisible by some other, smaller number can be represented as a rectangle. Here are just two examples of one number:

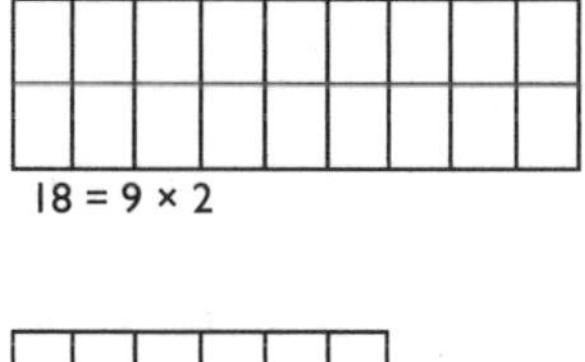

$18 = 9 \times 2$

$18 = 6 \times 3$

It is much more fascinating to find numbers that *cannot* be represented by a rectangle; to be more precise, numbers that are divisible only by 1 and themselves. For example, 17.

There is no way to arrange the squares into any rectangle other than the one below.

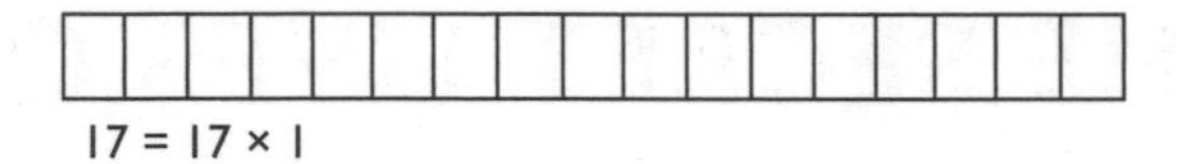

A number is termed a prime number if it has exactly two different divisors: 1 and itself. A number that is not a prime number is called a composite number.

Here are the first prime numbers: 2, 3, 5, 7, 11, 13, 17, 19, 23, 29, 31, 37, 41, 43, 47 … The list goes on. (If you carefully read the definition of a prime number, you will understand why 1 is not included in the list.)

Prime numbers are the building blocks for the entire population of numbers, since every composite number can be represented in one, and only one, way as a product of prime numbers, where any prime number may appear in the product more than once.

For example: $72 = 2 \times 2 \times 2 \times 3 \times 3 = 2^3 \times 3^2$

For people who don't consider themselves to be members of the community of mathematicians, the fact that every number can be represented in one and only one way as a product of its prime factors is perfectly obvious. But this fact is not perfectly clear to mathematicians – they have to go to the trouble of proving it. Please don't condemn mathematicians for being so punctilious; over the passage of time there have been so many things that seemed "perfectly obvious" but were eventually proved to be untrue that mathematicians have categorically decided that each and every assumption must be substantiated through proof. You may assume that a bunch of zeroes added together will give you zero, but as you will see later in this book, the sum of zeroes does not always end up to be zero, and if you can't trust zeroes, who can you trust?

But I digress. Let us return to the subject of prime numbers.

The first thing we might ask as we begin the relationship we are about to develop and nurture with prime numbers is: "How many prime numbers are there?"

The first to discover the answer to this question was the Greek mathematician and the father of the theory of geometry, Euclid. Anyone who has ever studied geometry – no matter where and no matter when – is familiar with Euclid. We have all learned Euclid's postulates (axioms) that only one unique line can pass through any two points, and that two parallel lines never meet. In fact, classic geometry bears his name and is called Euclidean geometry. And, even though Euclid developed this geometry more than 2,000 years ago, it is still being taught *exactly* the way he wrote it. Can you imagine biology or chemistry or physics being taught today using only the knowledge of more than 2,000 years ago or even 200 years ago?

Euclid's geometry has made a profound impression on the best minds of civilization, one of whom was the philosopher of philosophers, Baruch Spinoza. Spinoza was so taken by Euclid's methods of constructing geometry from axioms and basic concepts that he used this approach in his most important work, *The Ethics*. Of course, Spinoza was not discussing points and lines in his book. He was pondering the concept of God and man's place in his world. But the way in which he presents his arguments is pure Euclidean: Spinoza presents his basic concept, lays down specific axioms and then uses them to prove theorems. In fact, the original name of Spinoza's magnum opus is *Ethica Ordine Geometrico Demonstra.* (Although the English title is *The Ethics*, the exact translation of the Latin is *Ethics, Demonstrated in Geometrical Order.*)

Back to Euclid. Before we see his answer to "how many prime numbers are there?", let us think for a bit on our own.

First of all, we must determine if the number of prime numbers is finite or infinite.

If their number is finite, then what is the greatest prime number?

If there are infinite prime numbers, can this be proven?

Is it fathomable that some really huge, extraordinarily large number will not be divisible by anything other than 1 and itself and be declared a prime number?

Is there a formula that can be used to give all the prime numbers?

EUCLID'S THEOREM

There are infinitely many prime numbers.

I will present here two proofs. One is brief and emphasizes the beauty of Euclid's great idea. The other proof is essentially the same, but is longer and will serve to explain the shorter one in detail.

The short proof

Let us assume that 2, 3, 5, 7, 11, … P is the complete list of all the prime numbers up to some prime P.

Now, let us form a new number, S, where $S = (2 \times 3 \times 5 \times 7 \times 11 \times \ldots \times P) + 1$.

Now, S is either a prime number or it is divisible by (one or more) prime numbers *larger* than P. One way or another, P ends up not being the greatest prime number. Therefore the number of prime numbers must be infinite.

QED.

Are you convinced? If you are, you can skip the long proof; if not – read on!

The long proof

Here, as well, we shall assume the existence of the greatest number in the list of prime numbers, and then we shall prove that this situation is impossible, thus proving that the prime numbers are infinite. (This type of proof, where one begins with some assumption and then proves that the assumption is impossible is called, in mathematical terms, "proof by contradiction". This is a simple yet elegant concept that is totally natural for mathematicians, however many people feel some conflict with the idea when they first stumble upon it.)

If the number of prime numbers is finite, then it will be possible to find the greatest prime number, which we shall denote as P. Now, let us write down all the prime numbers: 2, 3, 5, 7, 11, 13, 17, … P.

We then form the following number: $S = (2 \times 3 \times 5 \times 7 \times 11 \times 13 \times 17 \times \ldots \times P) + 1$. That is to say that S is equal to the product of all the prime numbers in our list plus one.

So, what is S divisible by?

It is not divisible by two, since the expression within the parenthesis is an even number (2 is one of the factors in the expression). Adding 1 makes S odd.

S cannot also be divisible by 3. The reason for this is the same: the number represented in the parentheses is divisible by 3 (because it is a factor in the expression), therefore adding one to it means it isn't (actually S leaves a remainder of 1 when divided by any prime in the list).

S also cannot be divided by 4, since it is not divisible by 2. In fact, any number that is divisible by some divisor is also divisible by its prime factors. Any number divisible by 6, for example, is also divisible by 2 and 3.

If we continue in this fashion and think a bit more, we will come to the realization that S cannot be divided by 5, nor 6, nor 7, nor by any number up to and including the prime number P, which we have assumed to be the greatest prime number. This leads us to two possibilities:

1 S is a prime number greater than P, or
2 S is divisible by some prime number that is not included in the list, that is to say, a prime number that is greater than P (because we have already seen that it is not divisible by any of the primes less than or equal to P).

No matter which possibility we choose, we have reached a contradiction to our original statement, namely that P is the greatest prime number. If our assumption that P is the greatest prime number has led to a contradiction, this means that there is no "greatest" prime number, meaning the number of primes is infinite.

QED.

Important note: Euclid's proof is not especially constructive. That is to say, it does not provide a simple recipe for obtaining more and more prime numbers. S, as

By the way, in case you are wondering, QED comes from the Latin, *Quod Erat Demonstrandum,* meaning "what was to be demonstrated" and is the joyful notation that every mathematics student proudly scrawls at the bottom of his exercise when he has finally worked his way through some long complicated proof.

Spinoza used this Latin short form a lot. Interestingly, Euclid himself actually used the Greek initials *OEΔ*, which look similar and stand for *hoper edei deixai* ***(ὅπερ ἔδει δεῖξαι)*** – "what was to be shown".

we pointed out, does not necessarily have to be a prime number; it can be a composite number *divisible* by a prime number greater than P.

I shall demonstrate:

Let us assume that 3 is the greatest prime number in existence (an assumption that is, of course, as wrong as wrong can be). We form S, which is (2 × 3) + 1 = 7, and 7 is indeed a prime number. This is also true for S = (2 × 3 × 5) + 1, S = (2 × 3 × 5 × 7) + 1, and S = (2 × 3 × 5 × 7 × 11) + 1.

But then we get to an example of the second alternative:

(2 × 3 × 5 × 7 × 11 × 13) + 1 = 30,031 = 59 × 509.

In other words, (2 × 3 × 5 × 7 × 11 × 13) + 1 is a composite number divisible by the prime numbers 59 and 509, both of which are greater than 13, which temporarily appeared in the role of "greatest prime number". Therefore there is no contradiction whatsoever in Euclid's proof – it is flawless.

It is interesting to note that not a few people who are exposed to Euclid's proof feel that, had they not seen it, they would have managed to discover it on their own. (*Meh … multiplying prime numbers and adding one to the result. How easy is that?! It wouldn't have taken me more than a couple of minutes to figure it out.*) In most instances, this is an illusion. The simplicity of the proof demonstrates its beauty and genius.

I have met important mathematicians who are convinced that Euclid's proof regarding the infiniteness of prime numbers is one of the most beautiful theorems in mathematics ever. Were I an important mathematician, I would certainly add my vote to theirs.

Mersenne numbers and *The Guinness Book of Records*

The fact that there are infinite prime numbers guarantees that we will never be able to write the ultimate list of all prime numbers. There will always be another prime number that is larger than the largest prime on our list.

The number that holds the honourable title of "largest prime number ever discovered up to 2018" is $2^{77,232,917} - 1$. I don't recommend trying to calculate this number to write it in your notebook – your notebook will simply not have enough pages to contain the number. If you consider that the estimated number of atoms in the universe is less than 2^{320}, you can possibly begin to comprehend how stupendous $2^{77,232,917}-1$ is. This number has 23,249,425 digits, nearly one million (!) digits more than the previously recorded prime number, discovered in January 2016: $2^{74,207,281} - 1$ (a number with "only" 22,338,618 digits). In contrast, 2^{320} has a mere 96 digits. Everything is relative!

By the way, this monstrous number was proven to be a prime number not by a flesh-and-blood human mathematician, but by a server network project called GIMPS (Great Internet Mersenne Prime Search).

So, what is a "Mersenne" number? The correct question might actually be: "who" is Mersenne? Numbers in the form of $2^n - 1$ are called Mersenne numbers for French philosopher, theologian, musicologist, and mathematician Marin Mersenne (1588–1648). (If his list of titles is not sufficiently impressive, let me add another: he was the first person to measure the speed of sound.)

Are Mersenne numbers always prime? Absolutely not.

For example, $2^4 - 1 = 15$ is not a prime number ($15 = 3 \times 5$).

Those for whom their high-school studies are still fresh in their minds (or those who are still in school at present) probably know that if the value of its exponent is not a prime number, then that Mersenne number will not be a prime number. The reason for this is that in this case the number can always be divided into two factors. It works along the following lines, for which $2^6 - 1$ has kindly volunteered to serve as an example:

$$2^6 - 1 = 2^{2\times3} - 1 = (2^2 - 1)\,(2^4 + 2^2 + 1) = 3 \times 21.$$

In other words, if the exponent is a composite number, there will always be a way to factor the corresponding Mersenne number, thus proving it is also composite. The decomposition is given by the general formula:

$$2^{n \bullet m} - 1 = (2^n - 1) \bullet (1 + 2^n + 2^{2n} + \ldots + 2^{(m-1) \bullet n})$$

If the above formula doesn't really grab you, don't worry. In truth, the formula is not really important. What is important is the fact that if the exponent is *not* a prime number, the Mersenne number with that exponent will also not be a prime number.

If a composite exponent guarantees a composite Mersenne number, the next natural question is surely: "Does a prime exponent guarantee a prime Mersenne number?"

Let's try to check it out.

$2^2 - 1$, $2^3 - 1$, $2^5 - 1$, and $2^7 - 1$ are all prime numbers (3, 7, 31 and 127, respectively). So far, so good. The next prime number after 7 is 11, but $2^{11} - 1$ is *not* a prime number! $2^{11} - 1 = 2{,}047 = 23 \times 89$.

This is really a shame, but a prime exponent does not guarantee that the corresponding Mersenne number will also be a prime number. If it were, that would be a simple way to find more and more prime numbers. For example, we could take that colossal prime number we met a couple of lines above, use it as the exponent of 2 and then subtract 1, and we would have a new – and even more colossal – prime number. Such a number would have as its exponent a number with more than 20 million digits. Try to imagine how absolutely enormous this number would be – outside the bounds of imagination of mere mortals. Is this number actually a prime number? I don't know and I don't expect that I will ever know.

All the same, the fact that there are infinite prime numbers does not guarantee infinite Mersenne prime numbers. As of today, we know of 50 Mersenne prime numbers.

Here are the first eight: 3, 7, 127, 8,191, 131,071, 524,287, 2,147,483,647

The ninth number in the series is $2^{61} - 1$, which works out to a number with 19 digits, and I decided not to bore my computer screen by writing down these numbers – I think it has suffered enough.

Mersenne explored these numbers that bear his name today in a paper he published in 1644. It had the impressive title *Cogitata Physico-Mathematica* (Thoughts about Physics-Mathematics). He checked all the prime exponents up to 257 and concluded that $2^p - 1$ would be prime for P = 2, 3, 5, 7, 13, 17, 19, 31, 67, 127, 257. The correct list is a bit different and is actually the following: P = 2, 3, 5, 7, 13, 17, 19, 31, 61, 89, 107, 127.

You can decide whether you think Mersenne's rate of success is impressive or not.

Mersenne numbers and perfect numbers

Do you remember the perfect numbers we met in the section about Pythagoras? Just in case you have already forgotten, perfect numbers are numbers where the sum of their proper divisors is equal to the number itself. Euclid already knew that if $2^p - 1$ is a prime number, this number multiplied by 2^{p-1} would always give a perfect number. (Of course, Euclid didn't call such numbers "Mersenne" numbers. Not only had Mersenne not been born yet, the parents of the parents of the grandparents of his grandparents had not even met yet.)

I will demonstrate. $2^3 - 1$ is a prime number (7), therefore $(2^3 - 1) \times 2^2 = 28$, which is a perfect number. Similarly, $2^5 - 1$ is a prime number (31) and therefore $(2^5 - 1) \times 2^4 = 496$ is a perfect number. With the kind and generous assistance of the largest prime number known today, we can now construct the largest known perfect number: $(2^{77,232,917} - 1) \times 2^{77,232,916}$

I don't recommend trying to calculate the number and checking if this is true. I can assure you that all the factors of this monstrous number added together will yield the number itself. To paraphrase the words of the great German philosopher Immanuel Kant, I had therefore to remove knowledge in order to make room for belief.

Okay. The time has come to leave the subject of world records and commence developing some of those brain muscles.

Brain-twisters for readers with a mathematical background

1) Prove that if $2^p - 1$ is a prime number, then $(2^p - 1) \times 2^{p-1}$ will be a perfect number.
2) 28 is a triangular number.

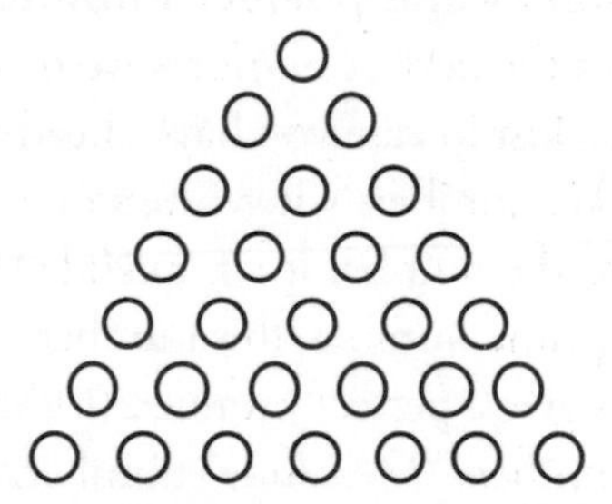

   Are all even perfect numbers triangular?
3) Famous Swiss mathematician Leonhard Euler (who we shall soon meet) proved that the converse is also true. That is to say, every even perfect number has the form $(2^p - 1) \times 2^{p-1}$ when both p and $2^p - 1$ are prime numbers.

You are welcome to try your skills, or, alternatively, find Euler's proof.[1]

The search for the miracle formula

Okay, we now understand that there are infinite prime numbers. The next logical question is whether there is any order to their appearance. Is there some formula that gives only prime numbers? Is there a formula that gives *all* the prime numbers? What would the formula for the number of prime numbers up to n look like?

We've barely parted from him and here we are about to again meet the great Swiss mathematician Leonard Euler (1707–1783).

In 1772, Euler noticed that the expression $n^2 + n + 41$ (any expression of the type $ax^2 + bx + c$ is called a quadratic polynomial) produces prime numbers as long as n is less than 40. For example, for $n = 0, 1, 2, 3, 4, 5, 6$, we get, respectively, the following values: 41, 43, 47, 53, 61, 71, 83. (Note that the differences between the values are 2, 4, 6, 10, 12.)

It is totally clear that Euler's formula cannot continue giving more and more prime numbers. Anyone who remembers even a smattering of the laws of mathematics from grade eight will know that if $n = 41$, the result will not be a prime number, since in this case, all three addends are divisible by 41, meaning that the sum will also be divisible by 41.

If we think a few moments more, we will also realize that the formula cannot give a prime number if $n = 40$. We shall write it thus:

$$40^2 + 40 + 41 = 40(40 + 1) + 41 = 40 \bullet 41 + 41$$
$$= 41 \bullet (40 + 1) = 41^2$$

Not only is the result not a prime number, it is actually a perfect square: 1,681.

Note that 1,681 has a very interesting property: it is the *only* four-digit number that is not only a perfect square, but each of its two parts, 16 and 81, are also perfect squares. (We ignore trivial examples such as 1,600.)

Remark: No quadratic polynomial $ax^2 + bx + c$ has ever been proved to generate infinitely many primes.

Dirichlet's theorem

When I was taking a course in number theory at Tel Aviv University, the lecturer, Professor Gregory Freiman, proved to us the following theorem:

The arithmetical series $an + b$ will produce infinite prime numbers provided that a and b are relatively prime to each other, that is to say, they have no common factors greater than 1.

The proof for Dirichlet's theorem (Gustav Lejeune Dirichlet, 1805–1859) is exceptionally beautiful, but it took the lecturer four lessons to explain it and it slides into fields of mathematics that are far beyond the scope of this book. Since I promised to use only the basic arithmetical operations, I shall just explain the statement of the theorem as simply as I can.

Let us choose two relatively prime numbers (that is, numbers that do not have any common factors), for example $a = 3$ and $b = 4$. (Remember, the numbers *themselves* needn't be prime numbers; they just need to be relatively prime to each other.) Our basic expression will be $3n + 4$. We calculate the series for consecutive values for n, beginning with $n = 1$.

The series of numbers obtained will be 7, 10, 13, 16, 19, 22, 25, 28, 31 …

Now, you have probably already observed that not all the values in the series are prime. But Dirichlet's theorem doesn't claim that they will be. Dirichlet's theorem states that infinitely many prime numbers will *appear* in this series, as will happen in any series where a and b are relatively prime. It is clear, of course, that infinitely many composite numbers will also appear in these series. For example, in the series $3n + 4$, whenever n is a multiple of 4, the result, of course, will be a composite number.

By the way, the source of the name Lejeune Dirichlet is an interesting story. Dirichlet's family hails from Richelette, a small hamlet near Leige, Belgium. He was called *"le jeune de Richelette"* – the young man from Richelette.

The kingdom of composites

Many years ago I was appointed to teach in a very special project in the framework of the School of Mathematics at Tel Aviv University. Professor Beno Arbel was responsible for discovering high-school students with exceptional talent in mathematics, and my job was to teach them a bit and prepare them for academic studies in parallel with their studies in high school. The main idea was for them to complete their bachelor's degree, or even their master's degree, before they finished high school or a short time thereafter.

I used to give them problems to solve that I got from my private collection of International Mathematics Olympiad because, in my opinion, tough problems really bring out the best. One problem I would give in the warm-up stage was the following.

Problem

Write 100 consecutive numbers so that they do not include any prime numbers.

At this point, you probably know what I am about to write. If you think that it is "try to think a bit before you read further", you are totally correct.

A tiny hint

This is not an easy exercise. The first thing that you will certainly think is that such a series of consecutive numbers must begin with a pretty large number – we already know that there won't be one hundred consecutive numbers without a single prime number among them in the lower numbers.

Keep thinking.

While you are thinking, I will take the opportunity to introduce you to (or remind you of) a very important symbol that simplifies writing and thinking. Obviously, there is a reason that I am presenting this symbol right now; it will help us to solve the problem. The symbol is the factorial symbol, which is the exclamation mark (!). In mathematics, $n!$ is defined as the product of all the numbers from 1 to n, that is $n! = 1 \times 2 \times 3 \times 4 \times 5 \times \ldots \times (n-1) \times n$.

For example, 5! is $1 \times 2 \times 3 \times 4 \times 5$. (Once, one of my students was absent from the lesson in which I introduced factorials. When he saw the notation 5!, he called it "five wow!") It is immediately apparent that 5! is divisible by all the numbers that make up the product. In other words, $n!$ is divisible by all the numbers from 1 to n.

For the sake of integrity, I have to point out the 0! is defined as 1 so as not to compromise the basic formula when defining $n! = (n-1)! \times n$.

Now give our problem a second try!

Any idea? If not, keep on reading.

A big hint

I hope that this time that we spent with factorials has brought you closer to a solution. You can be sure that factorials will be involved somehow. But how?

With which number should we begin? Maybe 100!? Nope, that's no good. The next number, 100! + 1, might very well be a prime number?

But, if … Do you see the solution?

A colossal hint

Maybe we should start with 100! + 2? This seems like a better idea. This number is divisible by 2 since both 100! and 2 are divisible by 2, and hence it is not prime. We are off in the right direction.

The next number, 100! + 3, is similarly divisible by 3, and if we continue thus … 100! + 100 is divisible by 100. Unfortunately, there is no way to instantaneously know if 100! + 101 is a compound number.

We were so close. But unfortunately, there are only 99 numbers from 100! + 2 to 100! + 100. Such a pity. A great idea down the drain.

Wait! Down the drain? No way! All it needs is a bit of tweaking.

The solution

We can start our series of numbers with 101! + 2 and end with 101! + 101. Now we have a series of 100 consecutive numbers that are, without any shadow of a doubt, composite numbers.

Clearly, one will now be able find a series of any length that will not include any prime numbers. For example, for a series of 1,000 consecutive composite numbers, we simply have to begin the series at 1,001! + 2. What this implies, of course, is that there will be fewer and fewer prime numbers the more we hit *really* large numbers.[2]

More about the frequency of prime numbers

As the numbers grow, the average difference between two consecutive prime numbers also gets larger. However, there is a theorem that puts an upper bound on the sparseness of the prime numbers among the natural numbers. It claims that the quotient $\frac{P_{i+1} - P_i}{P_i}$ approaches 0 as i approaches infinity where P_i represents the prime number in the i place.

I shall translate this mathematical lingo into language that non-mathematicians can understand. What the above theorem means is that the gap between prime numbers relative to the prime itself gets smaller as i becomes larger. Here is the list of values beginning with i =1 (Just to be clear, in the first row i is 1. Thus P_i is the first prime number, 2, and P_{i+1} is the second prime number, which is 3. In the second row *i* equals 2 and the primes are $P_2 = 3$ and $P_3 = 5$ and so on.):

$$(3 - 2)/2 = 1/2$$
$$(5 - 3)/3 = 2/3$$
$$(7 - 5)/5 = 2/5$$
$$(11 - 7)/7 = 4/7$$
$$(13 - 11)/11 = 2/11$$
$$\vdots$$
$$(103 - 101)/101 = 2/101$$
$$\vdots$$
$$(433 - 431)/431 = 2/431$$
$$\vdots$$
$$(3{,}539 - 3{,}533)/3{,}533 = 6/3{,}533$$
$$\vdots$$

As you can see, the expression $\frac{P_{i+1} - P_i}{P_i}$ has a tendency to get smaller and smaller as *i* increases (the expression does not get monotonically smaller, it just trends smaller as P gets larger) because as we get up into higher prime numbers, the numerator becomes *much* smaller than the denominator. This means that the *difference* between consecutive prime numbers (the numbers in the numerator) grows more slowly than the value of the prime itself, thus making the fraction smaller.

Although there is some instability early on in the calculations, if one considers the general trend, the gap between primes tends to get smaller compared to the size of the numbers themselves.

A direct path to a PhD

Despite years and years of research, there is more by far of things we *don't* understand about prime numbers than things we do.

Here are a few problems (out of many) that nobody, to the best of my knowledge and at this point in time, has solved. Perhaps you would like to try your hand at solving them. I promise you that if you do manage to solve even one, you will be instantly awarded a doctorate in mathematics and have your share of glory. (If you are still studying in school or university, solving these problems will give you an ultimate exemption from further attendance in classes or lectures.) That is the good news.

The bad news is really bad. It is not a coincidence that no one has yet managed to solve these problems. They are exceptionally difficult! It is hard to imagine the efforts that mathematicians have invested in trying to solve them. "There are no free lunches," our economists tell us. Well, I would have to add that "there are no luxurious meals without a high price".

Twins, triplets, cousins, and sexy primes

A pair of prime numbers are considered twins if the difference between them is equal to 2. For example, the pairs (3, 5), (5, 7), (11, 13) … (431, 433) … are twin primes.

Are there infinitely many twin primes?

Just because there are infinite primes does not mean that the answer to this question will be affirmative.

Prime triplets: A mini quiz

Here is a prime triplet:[3] (3, 5, 7). Prove that this is the only possible "triple twin" possible.

Prime cousins

Prime number pairs where the difference between them is 4, such as (3, 7), (7, 11), (19, 23) … (223, 227), are called cousins. Are there infinitely many pairs such as these?

Sexy primes

Prime number pairs that differ from each other by 6 are termed sexy primes. (Just look at what mathematicians consider sexy!) Here are a few of the sexiest pairs of the year: (5, 11), (7, 13), (11, 17), (17, 23), (23, 29) … (191, 197) …

Just look at the philandering going on here! 5's partner, 11, has also paired up with 17, who is flirting with 23, who is cheating with 29. But 29 remains faithful to 23. What a lot of potential material for a really awful romance novel!

Nobody knows if the numbers of twin primes, prime cousins, or sexy pairs are finite or infinite.

Note for mathematicians: The convergence of prime reciprocals

Let us examine the following series consisting of only twin primes:

$$(1/3 + 1/5) + (1/5 + 1/7) + (1/11 + 1/13) + \ldots + (1/857 + 1/859) \ldots$$

In 1915, Norwegian mathematician Viggo Brun proved a theorem that became famous and today bears his name. In his theorem, Brun showed that the aforementioned series converges and its sum is equal to approximately 1.9 (1.90216…).

Were the series divergent, we would know that there are infinite twin pairs. However, the fact that it converges does not tell us the slightest thing about the finiteness or infiniteness of twin pairs.

If we could prove that the value of the series cannot be expressed as a fraction – such numbers are called irrational numbers – the problem would also be solved, showing there are infinitely many twin primes (the sum of finitely many rational numbers is a rational number). However, the result is rational, which does not shed light on the issue of the infiniteness (or not) of twin primes. (For the non-mathematician, rational and irrational numbers will be explained shortly.)

The series for prime cousins is $(1/7 + 1/11) + (1/13 + 1/17) + (1/19 + 1/23) + \ldots + \ldots$ and the sum converges to approximately 1.197 (1.1970449…).

Stable primes

A prime number is termed "stable" if every arrangement of its digits also gives a prime number. For example, 199 is stable because 919 and 991 are also prime numbers. 13 is also a stable prime, since 13 and 31 are both prime numbers.

Run a computer program to search for stable prime numbers, and you will discover that after a trifling number of numbers (the last of which was 991), the only stable primes that appear are those that are composed of just the repeated digit 1. The first of these is 1,111,111,111,111,111,111.

Another open problem is the following: are there stable prime numbers larger than 991 yet not composed of all 1s? A tiny tip: the only numbers that can be included in any stable prime number are 1, 3, 7 and 9. Obviously, if any even number or the number 5 is involved, one of the arrangements will be a composite number.

Palindromes

A palindrome is text that can be read the same backwards as forwards. "I prefer pi" is an example of a palindromic sentence. Are there palindromic prime numbers? There are. In fact, there are quite a few: 919, 101, 14,741 … and many more excellent examples (the largest proven palindromic prime has almost half a million digits). However, what is not yet clear is whether their number is finite or infinite. Why don't you sharpen up your pencils, warm up your computer, and see what you can come up with.

The Legendre conjecture

The eighteenth-century French mathematician Adrien-Marie Legendre (1752–1833) proposed the conjecture that between n^2 and $(n + 1)^2$ there will always be at least one prime number.

Let us examine $n = 2$. Between $2^2 = 4$ and $3^2 = 9$, we have the prime numbers 5 and 7. Many mathematicians intuitively believe that this conjecture is correct, but as we have already stated, one cannot trust intuition alone when one is dealing with mathematics.

In the section entitled "The kingdom of composites", above, we learned that we can find a sequence of consecutive composite numbers (i.e. without any prime numbers) that is as long as we may desire. A student in one of my classes believed that this fact contradicted the Legendre conjecture, thus proving it wrong. He was wrong. One cannot form a series of consecutive numbers wherever we want. In our example, if you remember, the 100-number series only began at 100!. Now, 100! is an *enormous* number,[4] and in the area where this series of numbers resides, the gap between two consecutive squares is huge, theoretically leaving room for at least one prime number. Observe that the gap (that is to say, the difference) between 100! squared and (100! + 1) squared is the following:

$$(100! + 1)^2 - 100!^2 = (100!^2 + 2\times100! + 1) - 100^2 = 2 \times 100! + 1.$$

That, too, is one humongous gap!

At the time of writing, nobody has proven whether the Legendre conjecture is correct.

WOMEN IN THE WORLD OF MATHEMATICS

Up to now, most of the mathematicians we have met have been male. Have there been no important female mathematicians throughout history? There were. And how there were!

At this point, I will take a little break from prime numbers to tell you about some notable female mathematicians. It might be said that two of the greatest female mathematicians ever were Russian mathematician Sofia Kovalevskaya (1850–1891) and German Jewish mathematician Emmy Noether (1882–1935), a staunch admirer of Einstein.

It is impossible to be a mathematician without being a poet in soul.

(Sofia Kovalevskaya)

However, the history of female mathematicians begins much earlier.

In the ancient world

It is said that Pythagoras's wife, Theano of Croton, was a mathematician and physicist, and was also involved with medicine and psychology – a Renaissance woman before the Renaissance. Damo, the daughter of this first man-of-numbers, was also intrigued by mathematics, was a member of the Pythagorean sect, and quite likely contributed to at least some of the doctrines ascribed to her father.

Hypatia of Alexandria (born in the second half of the fourth century) was, without a doubt, the most famous female mathematician of the ancient world. Her father, mathematician and philosopher Theon, decided to raise her in the image of "the perfect human" and tried to bestow on her all the human knowledge accrued up to that point. He passed on to her all the learning that he had and, in addition, sent her to study in Athens and Rome. Most of her biographers point out that she surpassed her father in mathematical aptitude.

Hypatia was truly a woman of many talents. When she was in Alexandria, she studied Plato's and Aristotle's philosophy. She was also famous in her time as an astronomer and wrote a book with the impressive title *Astronomical Canon*, which is a set of tables describing the movements of the heavenly bodies. Hypatia was also renowned for her magnificent beauty, but according to biographers never married.

Hypatia was a pagan, something that her Christian neighbours were not amenable to. In 415, a gang of especially fanatic and violent monks accused Hypatia of religious sedition. They attacked her in the city square and then brutally tortured and murdered her.

Hypatia's story is so dramatic that it was just a matter of time until a movie based on her life would be made. Spanish director Alejandro Amenábar took up the challenge and made the movie *Agora* (2009). Naturally, the movie includes a love story.

Sophie Germain

This leads us to Sophie Germain, who is connected to the world of prime numbers and its myriad open problems.

Sophie Germain was born in Paris in 1776 (died 1831). Simon Singh, in his 1997 book *Fermat's Last Theorem*, tells us that at the age of 13 Sophie read that Archimedes would not abandon his efforts in mathematics despite threats to his life, this being the reason he met his death at the hands of a Roman soldier. The story so impressed Sophie that she decided that mathematics must be laden with beauty and interest if someone could be so engrossed in studying its secrets. (She would certainly have been just as awed had she known that Bertrand Russell thrice decided against suicide so he could learn just a bit more mathematics.)

Even though Sophie never learned mathematics formally and never received any academic title, she contributed considerably to the study of mathematics, especially in the fields of differential geometry and number theory. One of her more important contributions in the field of number theory was to reduce the number of possible solutions to Fermat's last theorem. Sophie Germain won a mathematics contest sponsored by the French Academy of Science, and she was the first woman allowed to participate in the Academy's seminars. Both a neighbourhood and a school in Paris are named after her, not to mention a crater on the planet Venus: the Germain Crater.

Sophie Germain primes

Now back to prime numbers and open problems.

A prime number, p, is termed a "Sophie Germain prime" if $2p + 1$ is also a prime number.[6] Here are a few examples of Germain primes: 2, 3, 5, 11, 23, 29, 41, 53, 83, 89 … For example, 5 is in this list because $2 \times 5 + 1 = 11$, and 11 is a prime number. On the other hand, 7 is not here, because $2 \times 7 + 1 = 15$ (not prime).

The wise reader will already have predicted the problem that no one has yet been able to solve: are there infinitely many Germain primes? Yes, that is the problem. However, one can also think of several other interesting problems.

(Take your time.)

Cunningham chains

Note the sequence of numbers 2, 5, 11, 23, 47. The number 2 is a Germain prime. Multiply it by 2 and add 1 and you get the prime 5, which is also a Germain prime that leads to 11, which is also a Germain prime that leads to 23, which is also a Germain prime that leads to 47. However, here is where the chain ends, because $47 \times 2 + 1 = 95$, which is not prime. Thus the series consists of four Germain primes plus one more.

Sequences of Germain primes of this sort are called Cunningham chains, after British military man and mathematician Allan J C Cunningham (1842–1928).

Here are some more problems:

- Are there longer chains? In fact, there are. My home computer is totally exhausted, but here is a humble example of a six-number chain that it turned up: 89, 179, 359, 719, 1,439, 2,879.
- Are there chains of any length?
- What would happen if we replace $2p + 1$ with $2p - 1$?
- Is there any point in examining $4p + 1$ or $4p - 1$?

Hah! It's really easy to *ask* difficult questions!

The Goldbach riddle, or, who wants to be a millionaire?

In 1742, several important events occurred. Johann Sebastian Bach composed the *Goldberg Variations* (there are hardly any real mathematicians who do not venerate this composition), poet Edward Young wrote *Night-Thoughts on Life, Death and Immortality*, the Indians rose up in Peru. And on June 7 of that year, the almost anonymous Prussian mathematician Christian Goldbach wrote a letter to the great Swiss mathematician (who we keep meeting over and over) Leonhard Euler.

To this day, Euler remains the most prolific mathematician ever. His writings number about 80 volumes of mathematical work in diverse fields. In contrast, Goldbach's height of glory is that he served as a tutor to Russian Czar Peter II (grandson of Peter the Great). Although one was Prussian and one Swiss, both Euler and Goldbach worked in the Saint Petersburg Academy of Sciences, which had been established by Peter the Great.

In his letter to Euler, Goldbach offered a conjecture that is today known as Goldbach's conjecture, and which is one of the oldest and best-known open problems in number theory and all of mathematics. It proposes that every even integer, beginning with 4, can be written as the sum of two primes (this is the modern version of the conjecture). For example, $4 = 2 + 2$, $6 = 3 + 3$, $8 = 3 + 5$ … Larger even numbers can often be written in more than one way as the sum of two primes. For example, $40 = 3 + 37 = 11 + 29 = 17 + 23$.

Let us examine the number 1,742, which is the year this conjecture was presented. Let's try, for example, $1{,}742 = 13 + 1{,}729$.

(Did you notice that 1,729 is the number of the taxi that Hardy took when he went to visit the ailing Ramanujan?!) So, by adding 1,729 to the prime number that represents dire bad luck, 13, we arrive at 1,742. There is only one major problem: 1,729 (as you must know already) is not a prime number, $1,729 = 19 \times 91$. Of course we can easily find other solutions, like $1,742 = 19 + 1,723$ or $1,742 = 43 + 1,699$ … check that now all the numbers are primes! You are welcome to propose your own solutions presenting 1,742 as a sum of two primes.

With all due respect to the many open prime-number problems presented up to this point, there is no doubt that the Goldbach conjecture is the most famous. The twin prime conjecture trails somewhere behind. All the other problems are remote compared to these two with respect to their familiarity and interest.

In 2000, Greek mathematics prodigy Apostolos Doxiadis's book *Tio Petrus e a Conjectura Goldbach* (Uncle Petros and Goldbach's Conjecture: A Novel of Mathematical Obsession) was published. The publisher, Toby Faber, offered a prize of one million dollars to anyone who could solve the Goldbach conjecture before April 2000. It was a masterpiece of marketing – maximum exposure with minimum risk – and, in fact, nobody claimed the prize. From now on, anyone who solves this problem will have to be satisfied with the much more modest (but much more exclusive) prize offered by Paul Erdős.

The solution to Goldbach's conjecture, when and if it arrives, may come from two different directions: either an even number that cannot be represented as the sum of two primes will be discovered (which you now already know is called proof by contradiction), or someone will prove why every even number can be represented this way. To date, a multitude of mammoth even numbers (up to 10^{18}) have been examined, and, so far, each and every one can be written as the sum of two prime numbers. Nevertheless, this doesn't mean a thing. Even if we check each and every even number up to 1,000,000,000,000,000! (that's 1 quadrillion *factorial*) and find that each and every one of those numbers can be written as the sum of two prime numbers, it is perfectly conceivable that the very next even number, 1,000,000,000,000,000! + 2, will be the first exception to our findings and will be the case that disproves the conjecture.

The Goldbach variation

Douglas Hofstader, in his book *Gödel, Escher, Bach: An Eternal Golden Braid*, suggests looking at the following variation of Goldbach's conjecture: is it possible to write every even number as the *difference* of two prime numbers? (Could you call this the Goldbach-Goldberg Variation?)

Let us begin: 2 = 5 – 3, 4 = 7 – 3, 6 = 11 – 5, 8 = 11 – 3 Obviously, some numbers have a few options: 10 = (41 – 31) = (29 – 19) = (23 – 13) = (17 – 7) = (13 – 3)

Despite the pronounced similarity between these two problems, there is an intrinsic difference. For the original version of Goldbach's conjecture, we can run a computer program for every even number, and it can check if two prime numbers added together will give that value in a finite amount of time. Even if the number is of great magnitude, we can be certain that, even though we may not be around to witness the event, the program will terminate at some point. On the other hand, there is no certainty that the computer will ever reach the end of its computations with the second version. Let us take an arbitrary number, say 2,010. It is absolutely uncertain when (and if) the computer will reach the end of the process, because, even if we check each and every prime number up to, say, 12,345,678,910 and do not find a pair of primes whose difference is 2,010, it doesn't mean that we won't find such a pair in the future. (I used 2,010 just to illustrate the idea. In reality, the computer won't have to work very hard to discover that 2,010 can be written as the difference of two primes, such as 2,017 - 7, 2,029 - 19, 2,039 - 29, and others.) In any event, this is very different from checking whether we can represent 2,010 as the *sum* of two prime numbers (which you already know is possible – the simplest of several options being 2,003 + 7).

The difference is this: to find the answer with respect to a sum, there are finite possibilities: one only has to check all the prime numbers up to the desired number itself. In the case of 2,010, we only have to examine all the primes up to 2,007 (the highest prime before 2,010). Even if, instead of 2,010, we check out 2,010!, this is still a finite number and the program will eventually reach its conclusion within a finite amount of time (longer than one can imagine, but still finite).

But to answer with respect to a difference, there are infinite numbers greater than the number we are seeking the answer for, therefore, we don't have a bound for the number of differences we may need to check and it may also happen that the process will never end.

Hardy compliments Fermat

Pierre de Fermat (1607–1665) discovered something interesting that is associated with prime numbers and is called Fermat's Christmas theorem. He showed that any prime number of the form $4n + 1$ (e.g. 5, 13, 17, 29, ...) is the sum of two squares, and every prime number of the form $4n - 1$ (e.g. 3, 7, 11, 19, ...) cannot be represented as the sum of two squares. Every prime number other than 2 is either of the form $4n + 1$ or $4n - 1$ (prove this to yourself).

For example, 41 is a prime number of the form $4n + 1$ $(4 \times 10 + 1)$, and it can be represented as the sum of two squares $(5^2 + 4^2)$. On the other hand, 19 is of the second form $(4 \times 5 - 1)$ and *cannot* be represented as the sum of two squares. While it is easy to show that 19, for example, is not the sum of two squares, proving Fermat's Christmas theorem in general is not simple.

In his book *A Mathematician's Apology*, G H Hardy concludes that Fermat's above discovery is an example of "elegant mathematics" and the most beautiful mathematical theorem alongside Euclid's proof regarding the infinity of prime numbers.

Well, on the subject of "conclusions", the time has come to conclude this section on the secret life of prime numbers and venture forth to the (boundless) world of infinity.

Mathematics, rightly viewed, possesses not only truth, but supreme beauty – a beauty cold and austere, like that of sculpture, without appeal to any part of our weaker nature, without the gorgeous trappings of painting or music, yet sublimely pure, and capable of a stern perfection such as only the greatest art can show.

(Bertrand Russell)

LESSON 4

PYTHAGORAS'S GREAT DISCOVERY

The origins of the mathematical theory of infinity, like almost everything else in Western civilization, stem from ancient Greece. It is interesting that the Greek word for infinity, ἄπειρον (*apeiron*), has two meanings. One meaning is an unbounded item; the other is more negative in nature: "an undefined thing." It was Anaximander, a sixth-century BC philosopher and astronomer, and Thales's pupil as well as Pythagoras's teacher, who was the first to introduce the concept of infinity into philosophy. In his cosmology, Anaximander saw infinity as one of the foundations of the world, a sort of unbounded, undefined embryonic material that is the source of all things. Some pre-Socratic philosophy scholars view Anaximander as the first metaphysicist to introduce an abstract concept of god into Greek philosophy.

One way or another, there was no place for *apeiron* in Pythagoras's world. You remember, of course, that Pythagoras was convinced that the world was built of numbers, and that everything in the world could eventually be represented using natural numbers, that is to say positive integers. The natural numbers were, in essence, Pythagoras's atoms.

It was Pythagoras himself who discovered that he was wrong.

An irrational number!!!

Ironically, the stumbling block to Pythagoras's reasoning that everything could ultimately be represented with the generous help of natural numbers rolled across Pythagoras's path as a result, believe it or not, of geometry, when he discovered that it is impossible to present the relationship between a square's side and its diagonal as a ratio of natural numbers.

I shall explain.

Let us begin with a square whose sides are one unit in length. We denote the length of the diagonal by *c*:

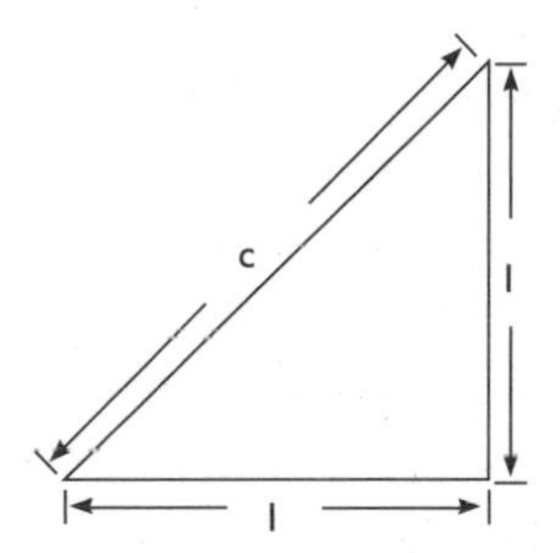

And thus goes the theorem that made Pythagoras famous: In a right-angled triangle, the square of the hypotenuse is equal to the sum of the squares of the other two sides.

Looking at our diagram, this means that $1^2 + 1^2 = c^2$, and therefore $c = \sqrt{2}$.

Note that $\sqrt{2}$ is simply a symbol for a number that when multiplied by itself gives us the value 2. Theoretically, we could have drawn a flower and say that it represents a number that, when squared, equals 2. Clearly, there is no integer whose square will be 2. (1 squared is 1 and 2 squared is 4, and there are no integers between 1 and 2.)

But is it possible that there is a fraction, $\frac{a}{b}$, that when squared gives 2? At this point, I will remind you that a number in the form of a/b, where a is an integer (including zero) and b is a natural number (a positive integer), are called rational numbers. Pythagoras certainly would have been very pleased had such a fraction existed, as this would have coincided nicely with his philosophy that everything in the world can be represented by natural numbers.

Extremely bad news was in store for Pythagoras!

We shall now prove that $\sqrt{2}$ can never be represented by a fraction of the sort a/b where both a and b are natural numbers. In other words, we shall prove that $\sqrt{2}$ is not a rational number.

We shall do this using proof by contradiction, which we have already encountered in this book. In other words, we shall first assume that what we wish to prove is false, that is to say that there exist two numbers, a and b, such that $-$ = $\sqrt{2}$. We shall then demonstrate that the logical outcome of this assumption leads to a contradiction.

We begin our proof by assuming that a/b is a reduced fraction, that is to say the fraction is written with the lowest possible denominator (so a fraction such as 21/14 or 15/10 would be written as 3/2). It is enough to show that there is no reduced fraction equal to the square root of 2, to show that $\sqrt{2}$ is an irrational number. This extra assumption about the fraction will be useful for the proof. It is a legitimate assumption, since every fraction can be written in its reduced form, and therefore if there is no *reduced* fraction equal to $\sqrt{2}$, then there is no fraction at all that is equal to $\sqrt{2}$.

So, we have our reduced fraction, a/b, and now we will assume that $a/b = \sqrt{2}$. A little rearranging gives us $\sqrt{2} \bullet b = a$, and then squaring both sides of the equation gives us $2b^2 = a^2$. This clearly shows that a^2 is an even number, meaning that a must also be even. Hence we can substitute $a = 2k$ into the previous equation to get:

$$2b^2 = (2k)^2$$
$$2b^2 = 4k^2$$
$$b^2 = 2k^2$$

We see that b^2 is an even number, meaning that b is also even.

However, if both a and b are even, this means that a/b is not a reduced fraction since both numerator and denominator can be divided by 2. Therefore this contradicts our initial assumption, which was that we began with a reduced fraction. In other words, we have just proven that $\sqrt{2}$ can never be the quotient of two integers. Conclusion: $\sqrt{2}$ must be an irrational number.

QED.

But what is the significance of what we have just proven?

The geometric interpretation means the following: we can easily construct a right-angled triangle with sides of unit length, and, with the same ease, construct its diagonal, but we cannot measure the length of this hypotenuse in relation to the sides of the triangle in a finite number of steps.

Such a simple geometrical concept – the hypotenuse of a triangle – negates the basic principle of Pythagoras's philosophy, which holds that everything is made up of *natural* numbers. It is easy to imagine that alongside the joy of his discovery, Pythagoras felt great disappointment.

Alternatively, we can use a calculator. Punch in $\sqrt{2}$ and see what happens. I punch it in and the result I get is

1.4142136. Now, use the method of long multiplication to multiply this number by itself. If this number is the *exact* square root of 2, multiplying it by itself should yield *exactly* 2. But it doesn't! (Check it yourself if you feel so inclined.) The reason that we don't get 2 is because the calculator only gives us an *approximation* of $\sqrt{2}$. Even if we purchase the best, top-of-the-line calculator that will give a greater number of digits after the decimal point, the result will still only approximate $\sqrt{2}$, but it will never be *exactly* $\sqrt{2}$.

If I switch from a calculator to a computer, I get the following result:

1.41421356237309504880168872420969807

If you're looking for an unbelievably boring activity for a rainy evening, feel free to try your hand at multiplying this number by itself and see if you get 2. You won't. Again, you will get something really close to 2, but not 2.

Time for explanations

A good way to understand what an irrational number is, is as follows: when we write that $\sqrt{2}$ is equal to 1.4142135... there is no easy way to explain what that ellipsis (...) represents. The irrationality of a number implies that 1) its decimal development is infinite and 2) there is never a repeating pattern.

When the number of digits after the decimal point is finite, the number itself is clearly rational – that is, there is no problem writing it as a fraction *a*/*b*. For example, 0.174271 = 174,271/1,000,000.

A number that has an infinite decimal development with a repeating pattern is also rational, although this might be a little harder to comprehend. Let us look, for example, at the number $r = 0.123123123123\ldots$ This number has a simple repeating pattern, and it is easy to prove that it is a rational number, that is to say, it can be written as a/b.

We multiply r by 1,000 (the number "1,000" was chosen because of the length of the repeating pattern) and subtract r from the result.

$$1{,}000r - r = 999r = 123.123123123\ldots - 0.123123123\ldots$$
$$= 123$$

Hence, $r = 123/999$, which can be reduced to the fraction 41/333, which is clearly the ratio of two integers, and if you calculate this using long division, you will see that, indeed, $41/333 = 0.123123123\ldots$

However, we cannot try a similar trick for $\sqrt{2}$ because its decimal development is infinite and non-repetitive. We can find fractions, 577/408 for example, that come pretty close to $\sqrt{2}$. This is a pretty good approximation, but it is just that – an *approximation*. It is interesting to note that Pythagoras himself refused to see $\sqrt{2}$ as a number. Many scorned him, and still scorn him, for that, without justification in my opinion.

It is important to remember that $\sqrt{2}$ is nothing but a symbol for a number that when multiplied by itself gives 2. As I have already pointed out, we could have symbolized this number with a flower instead of $\sqrt{2}$, and said that the flower represents a number that equals 2 when squared. Is there a difference between the conventional symbol and a flower? Maybe we should start using lots of flowers in mathematics and make it much more cheerful?

The only difference between the accepted symbol and our flower is that the flower is less convenient to use. And, in fact, we are not interested in symbolizing anything; we want to write the number whose square is 2. But this is a mission impossible, because no matter how many digits we write after the decimal point, it will never be enough. We would need to write infinitely many digits, and this can never happen.

Theodorus (465–398 BC), who was born about 30 years after Pythagoras died and was Plato's private mathematics tutor, proved that the square roots of 3, 5, 6, 7, 8, 10, 11, 12, 13, 14, 15 and 17 were also irrational numbers. Plato admired Theodorus and even mentioned him and his discovery about the irrationality of the square roots in his dialogue "Theaetetus".[1] Opinion is divided regarding the reason that he stopped at 17. In the dialogue, Theaetetus simply tells Socrates that Theodorus stopped there. The popular opinion is that Theodorus drew triangles in the spiral pattern that today bears his name. Continue the pattern, and you will immediately see the reason why his series of numbers stopped where it did. The spiral of Theodorus:

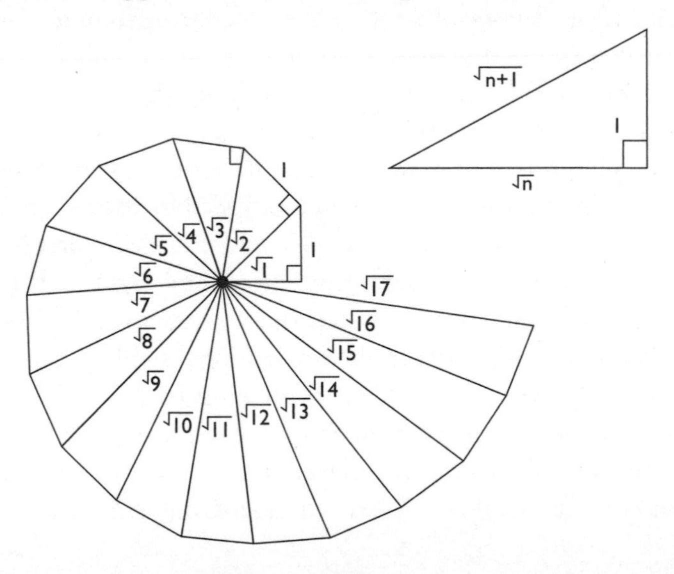

Brain-twister

1. Prove that the square root of 3 is an irrational number.
2. Try to prove that the square root of any integer is either an integer or an irrational number. (In other words, the square root of any integer other than a perfect square – 4, 9, 16, 25 … – will always be irrational.)

Okay. When Pythagoras decided that there was no number in existence whose square is 2, he exaggerated a bit. There *is* a number whose square is 2, and this number is an irrational number. Today, mathematics knows how to deal with such numbers without too much difficulty, despite our inability to write them down completely. Credit for the foundation of the mathematical theory of irrational numbers must be given primarily to three mathematicians – Richard Dedekind (1831–1916), Karl Weierstrass (1815–1897), and Georg Cantor (1845–1918). Make no mistake about it: it is not an easy thing to manage these numbers. Think, for example, how one would go about adding together $\sqrt{2}$ and $\sqrt{3}$ – both of which have an infinite decimal development.

How, indeed, do we add together

1.41421356237309504880168872420969807...

and

1.73205080756887729352744634150587236...?

The basic rules of addition that we learned at school tell us that we must start by adding the rightmost digits. But we cannot locate such a thing – the decimal development is infinite! What do we do? I told you not to mock Pythagoras for his refusal to accept irrational numbers as numbers.

The discovery of irrational numbers by Pythagoras is seen by many to be the most important mathematical discovery ever in the history of mathematics.[2]

Legend has it that Pythagoras asked his students to keep secret his discovery of the irrationality of the diagonal of the square with respect to its sides, yet one of them, Hippasus, betrayed his confidence (it is not clear whether for academic or political reasons) and revealed it publicly. The legend continues to say that Hippasus was evicted from Pythagoras's society, and some even say that he was drowned at sea (he simply did not return from one of his voyages around the Greek islands). Another version has it that it was actually Hippasus himself who discovered irrational numbers, and Pythagoras was actually not even involved.

More than two thousand years after the death of Pythagoras, Cantor would show that "almost" all real numbers are irrational. (These include two of the most significant numbers in mathematics: Euler's constant, e, and the ratio between a circle and its diameter, π.)

A comment and five exercises

I promised that I would not use anything in this book but the four basic mathematical operators, but what good is a "law" if it can't be violated at least once? That time has come now.

The numbers for which proof of irrationality is easiest are generated by the logarithm operation.[3] As an example, let us examine $\log_2(3)$, which is the base 2 logarithm of 3. Let us prove its irrationality. We begin by assuming that the fraction *m*/*n* is equal to this number:

$$\log_2(3) = \frac{m}{n}$$

Based on the definition of a logarithm and the laws of powers, it thus follows that $2^{m/n} = 3$, and $(2^{m/n})^n = 3^n$ and therefore $2^m = 3^n$.

However, it is impossible that any power of 2 can equal a power of 3: powers of 2 are always even and powers of 3 are always odd. Thus we have reached a contradiction. In other words, there are no numbers *m* and *n* for which

$$\log_2(3) = \frac{m}{n}$$

meaning that *m*/*n* cannot be rational. $\log_2(3)$ must be an irrational number.

Five brain-twisters

1 Prove that the golden ratio,[4] φ, is an irrational number.
2 The symbol of the Sect of Pythagoras is a pentagram within a pentagon.

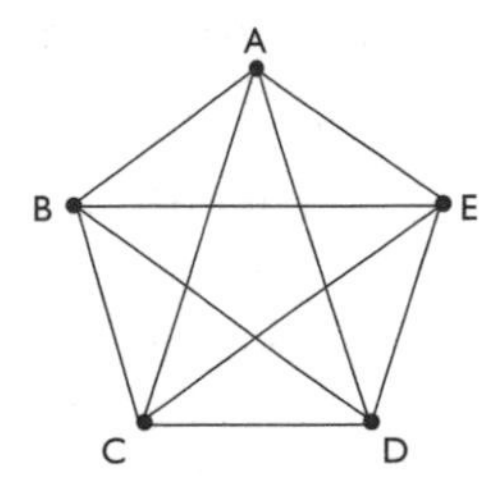

Prove that the ratio between the length of a diagonal of a regular pentagon to one of its sides is irrational. Moreover, show that it is not merely *any* irrational number, but the value φ (see above problem). That is to say, the ratio between any diagonal and any side of this figure is equal to the golden ratio!

$$\frac{AC}{AB} = \varphi$$

How lucky was Pythagoras that he never knew that those rogue irrational numbers, which he refused to accept as equal members of the family of numbers, were hidden within his iconic symbol?

The Pythagoras symbol can be upgraded thus:

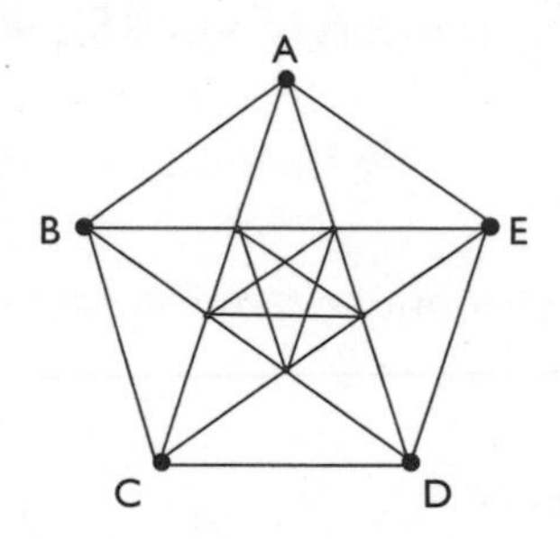

And this "upgrade" can continue ad infinitum!

3 Is the number 0.07007000700007… rational?
4 Is the following number rational: 0.123456789101112…?
5 What about this number 0.01123581321345589144…, which is generated by the Fibonacci's sequence 0, 1, 1, 2, 3, 5, 8, 13, 21, 34, 55 … (the sequence starts with 0 and 1 and each following number is equal to the sum of the preceding two numbers)?

LESSON 5

THE TORTOISE, ACHILLES AND THE ARROW – ZENO'S PARADOXES

The ancient Greeks' fear of the infinite was remarkably expressed in the famous paradoxes of Zeno of Elea, born circa 490 BC. His work culminated in Greece circa 450 BC. Not much is known about Zeno's life. Apparently, he spent most of it in his hometown of Elea, although Plato tells us in his dialogue "Parmenides" about a fascinating meeting held in Athens that was attended by Zeno, Parmenides, and the young Socrates.

Zeno's book (he apparently wrote only one) has not survived, and most scholars rely on Aristotle, who relates Zeno's paradox in his book *Physics*.

Zeno based his paradoxes on the philosophy of his teacher and friend Parmenides, so before we get to Zeno's paradoxes, let us learn about Parmenides and his unusual philosophy.

Parmenides' take on life

Parmenides, Zeno's teacher and friend (Plato even implied that Zeno and Parmenides were lovers), is considered an exceptional phenomenon not only in Greek philosophy, but also in the entire history of Western philosophy. The single known work of Parmenides is *On Nature*, written as a poem, which has survived only in fragmentary form. Parmenides describes two views of reality – "the way of truth" and "the way of senses". In "the way of truth" he explains that reality is timeless, uniform, infinitely dense, and unchanging. In "the way of senses", he explains the world of appearances and opinions, which is false and deceitful. Parmenides' philosophy of life was that the world of the senses is nothing but illusory, and the true universe, which can be known only by rigorous thinking, is silent and frozen in place. He claimed that the true universe is in exactly the same state now as it was a second ago, or a year ago, or a billion years ago, and will remain so for ever.

What??? It sounds weird even for a Greek philosopher.

Here is the essence of his philosophy and how he came about it.

Parmenides sought for a truth so obvious that any doubt regarding its veracity would be impossible. He wanted to base all his philosophy on this unquestionable truth. In mathematics, such a truth is called an axiom. After days of demanding contemplation, Parmenides fell asleep, and in his dream, Athena, the goddess of wisdom, Zeus's daughter and the patron of the city of Athens, helped him to find what he was looking for: *What-is is and what-is-not is not.*

Parmenides claims here that what exists is and what doesn't exist is not. This claim seems obvious and harmless. Sure? Keep on reading.

Armed with this axiom, Parmenides went on to propose

additional truths (in the language of mathematics, these are called theorems). The next thing that Parmenides comes up with is – **Parmenides' first theorem:** That which exists (what-is) is uncreated and imperishable.

The proof of this theorem is almost instantaneous – in mathematics we call this (as we have mentioned elsewhere) "proof by contradiction": we assume the opposite – "that which exists" was created at some point – and then test to see if it leads to a logical contradiction. If it does, this means that the initial assumption is incorrect.

If "that which exists" was created, it must have been created from something either "that is" or "that is not". Nothing can be created from something "that is not", because there is nothing there. But if "that which exists" was created from something "that is" it means that "that which exists" already existed. Therefore, "that which exists" was never created.

QED.

I will leave it to the clever reader to similarly prove that "that which exists" can never disappear.

The next thing that Parmenides proved was – **Parmenides' second theorem:** All that exists is uniformly and infinitely dense.

The proof for this is, again, amazingly simple. That which exists must be infinitely dense, for if not, this means that it has within it at least some of "that which does not exist". But "that which does not exist" does not exist. Parmenides' view that "what-is-not" is not, was time and again interpreted as his denial of the existence of void. Void is "what-is-not" and therefore non-existent.

Similarly, all that exists must be uniformly dense because if there is something that exists and is less dense than something else, it means that it has more of "that which does not exist". But "that which does not exist" does not exist!

Again, QED.

Are things starting to get weird? Wait and see what the cherry on the cake is!

Parmenides' great theorem: Motion is an illusion

The proof to this is immediately obvious – if everything is infinitely dense, how can motion be possible?

The fact that few would deny motion exists didn't concern Parmenides at all. He was not interested in the world of senses and opinions in which the possibility of error exists. In his world of truth (a world of axioms and theorems), everything stands still, nothing is created, and nothing disappears.

Paradox 1: The dichotomy, or the illusion of motion

Back to Zeno. The one book he apparently wrote was used to try to defend his mentor's philosophy. In particular, he wanted to corroborate Parmenides' claim regarding the impossibility of motion, the claim for which Parmenides was criticized – one might even say ridiculed – the most. (Indeed, it is a very strange thing to believe in the non-existence of motion.)

As part of Zeno's defence of his master, he introduced his famous paradoxes, which have fascinated countless mathematicians and philosophers for over two thousand years. Those who have invested serious brain time attempting to understand them include Aristotle, Maimonides, Descartes, Leibniz, Spinoza, Bergson, Russell, Lewis Carroll, Kafka, Sartre, Hegel, and Lenin (who read about the paradox in Hegel's book and wrote in his "Philosopher's Notebooks" that they aren't bad at all). There are many more.

So what are these paradoxes?

The first paradox is called "The dichotomy", in which Zeno uses an extremely rational and logical explanation to show that motion is impossible.

Look at the diagram below. Zeno claimed that in order to get from point A to point B, one must first pass the halfway point to reach C.

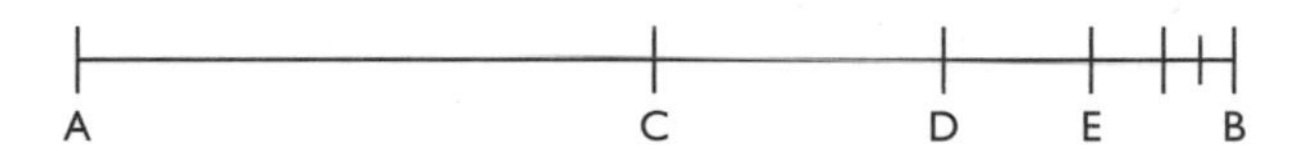

But from C to B, one must pass half of the remaining distance and get to point D. But getting there will not be any reason to rejoice, since now to get to B we must pass half the remaining distance and get to point E. And so on and so forth.

Zeno's argument: "It is impossible to pass infinite points in a finite time. Therefore it will be impossible to ever get from point A to point B." (Finally, we understand how to solve the problem of "how much time will it take a train to arrive at one point after it has left another point?" The answer is that it will never arrive.) Since points A and B are entirely arbitrary, it follows that it is impossible to get from anywhere to anywhere else. Hence, motion is impossible.

Many books offer a simple resolution to this paradox. The resolution goes (Zeno would have emphatically opposed the use of the word "goes") more or less like this: we assume that the time needed to travel a specific distance is proportional to its length. We can now prove that Zeno was wrong, because it is possible to travel infinite "half-distances" (which are getting smaller and smaller) in a finite time. For example, if we assume that it takes exactly one unit of time, say a minute, to get from point A to C, it will take half a minute to get from point C to point D (because that distance is half the previous one), and a quarter of a minute to get from point D to point E, and so forth. We denote the total time required to get from A to B by S.

$$S = 1 + \frac{1}{2} + \frac{1}{4} + \frac{1}{8} + \ldots.$$

If we divide both sides by 2, we obtain:

$$1/2\ S = \frac{1}{2} + \frac{1}{4} + \frac{1}{8} + \frac{1}{16} + \ldots.$$

We subtract the first equation from the second, and get:

$$S - ½\ S = \left(1 + \frac{1}{2} + \frac{1}{4} + \frac{1}{8} + \ldots\right) - \left(\frac{1}{2} + \frac{1}{4} + \frac{1}{8} + \frac{1}{16} + \ldots\right).$$

Hence,

$$½\ S = 1 \text{ and } S = 2.$$

In other words, it will take precisely two minutes to travel infinite half points to get from A to B.

Frankly, this solution is not so surprising. We began by assuming that it takes one minute to get from A to C, which is exactly half the distance, and therefore one cannot be surprised that the entire distance will take two minutes.

Zeno would most certainly have objected vociferously to this solution, because we assumed what needed to be proven. When we said, "If we assume that it takes a minute to get from A to C", we assumed that motion is possible and that we can get from A to C. However, that assumption is what needs to be proven, and this leads to what is called a circular proof.

Zeno would have probably clarified his standing by saying something similar to this: "How and on what grounds do you *assume* that you can get from point A to point C? It is totally obvious to me that you are making a monumental error. It's as clear as the sun on a summer day on a beautiful Greek island, that before arriving at point C from point A, one must travel half that distance and reach point X. He then must travel another half of the remaining distance to C to reach point Y, and so on and so forth. You certainly realize that it is impossible to pass infinite halfway points in finite time. Therefore it is impossible to get from point A to point C, which is in contradiction of what you just said. I trust that you understand that it is also impossible to get from A to X, because here, too one must pass half the distance, and blah, blah, blah. The truth is that it is impossible to get from any point to any other point. In other words, it is impossible to even begin any sort of motion whatsoever."

It seems that Zeno has won the day. Doesn't it?

What do you think?

Paradox 2: The heel of Achilles and the stealthy tortoise

Zeno's second paradox is, so it seems, his most famous. This paradox asserts that if a race were to be held between Achilles, the glorious fleet-footed hero of the Trojan War, and a tortoise that is not especially swift, as all tortoises are, and if the tortoise would be granted any head start whatsoever (even the most minuscule), Achilles would never be able to catch up to the tortoise. Does this sound illogical? Let us give Parmenides' favourite pupil the right to voice his argument.

Zeno explains it thus: "As soon as Achilles reaches the point where the tortoise began the race, he will discover that the tortoise is not there anymore. It is true that the tortoise only progressed slightly, but it makes no difference. Achilles will still be lagging."

Now, we simply have to repeat this argument over and over. Every time Achilles reaches the point where the tortoise had been, he will find that the tortoise has moved on. The tortoise progresses at the rate that a tortoise progresses, but Achilles can never catch up.

I think it is worth pausing here a moment (or even two) to consider Zeno's justification.

Let us assume that the race is 100 metres long. Also, let us assume that Achilles' speed is 10 metres per second (he's a well-known athlete and warrior, right?) and the tortoise shuffles forward at 1 metre per second (pretty fast for a tortoise). To make the competition fair, the tortoise will get a head start of 10 metres.

It is clear to us that Achilles will win the race, and easily. Our hero will have traversed those 100 metres in an impressive 10 seconds. The tortoise needs to travel only 90 metres, but because it takes him an entire second for each and every metre, he needs 90 seconds to cover the distance. To recap: Achilles gets to the finish line first, receives his laurel wreath, bows to his audience of fans, and waits with great patience for his rival to arrive. The tortoise, dripping with perspiration and on the verge of collapse, arrives a full 80 seconds after Achilles has finished.

This seems very obvious. But Zeno's take on the matter is totally different. Here is Zeno's argument.

As soon as Achilles hits the 10-metre mark, where the tortoise started out, he will discover that the tortoise is not there anymore since it has managed to toddle forth one metre and is now at the 11-metre mark. The distance between the competitors has been reduced from 10 metres to only 1 metre, but the tortoise is still in the lead.

When Achilles reaches the 11-metre mark, he will again discover that there is no tortoise waiting for him. In the time it took for Achilles to go from the 10-metre mark to the 11-metre mark, the tortoise has advanced 10 centimetres. (The tortoise is "running" at one tenth the speed of Achilles, so in the time that Achilles covers a distance of x, the tortoise covers a distance of $x/10$).

Here is a table showing the progress of our two "athletes":

Tortoise	Achilles
10	0
11	10
11.1	11
11.11	11.1
11.111	11.11

We learn from the table that the distance between Achilles and the tortoise constantly diminishes, yet the tortoise always retains some advantage. Furthermore, we can see that not only will Achilles never catch up to the tortoise, he will never even reach the 12-metre mark.

"What a bunch of baloney!" I hear you saying. "After just two seconds of the race, Achilles will be at 20 metres from the kickoff point and will lead the tortoise with certainty." This seems absolutely and perfectly apparent. Nevertheless, let us give Zeno the right to respond.

ZENO'S APOLOGY

Look, I'm afraid you haven't understood anything of what I was trying to explain. I offered a very convincing argument proving that Achilles will be never be able to catch up to the tortoise if that chelonian gets even the tiniest advantage at the start of the race. Are you trying to tell me that after two seconds Achilles will have reached 20 metres from the starting line, and will lead the tortoise by 8 metres, who is only 12 metres away from the start?

First of all, I didn't ask for any explanations. I just asked you to point out any error in *my* argument. That's not what you did.

Second, although you really didn't conduct yourself properly, I propose synthesizing our two arguments – mine and yours. The one possible conclusion that does not contradict any allegations made to this point is that two seconds can never pass. I have definite proof of this claim.

You see, in order for two seconds to pass, half that time first has to pass, that is to say, one second. But before that, half of that second must pass, and before that half of that half second (in other words, a quarter of a second), and then again half of the remaining time, and then again and again…

There is no way on earth that an infinite number of half-times can pass. Therefore, time does not pass, meaning it also does not exist. You are simply too immersed in the deceiving world of the senses.

Now, mind you, I don't believe in the act of reading – this is another widespread illusion in the world of senses – but nevertheless, once I read a few lines that almost make sense:

Time does not pass,
we pass. Yes we.
We do not waste any time,
time is what wastes us.

And if we are already talking about books, here is a person who wrote not a few books and read even more than he wrote, and he brilliantly defends my virtuous position.

His name is Bertrand Russell, an English philosopher and mathematician. (Yes, I realize he will live over two thousand years after time has done wasting me, but…)

Russell is considered one of the greatest thinkers of the twentieth century, and he had his own version of my famous paradox: Achilles and the tortoise (and incidentally, time also wasted Russell). Russell wrote his variation of my paradox in his article "Mathematics and Metaphysics", in which he also proffered on me the title of "father of the philosophy of infinity" – a title I no doubt find quite impressive, despite my custom of doubting everything, including my ability to doubt.

Some people claim that Russell's version is more sophisticated than mine, and that it isn't as easy to disprove. I don't think it's fair to compare versions – Russell came up with his version standing on my shoulders. Standing on his father's shoulder does not make a child taller than the father. I do believe, though, that not only is it not easy, but it is impossible to refute his version (or my version).

So said Russell: "Let us assume that the tortoise starts the race at some advantageous position ahead of Achilles. At any instant, the tortoise is at some particular spot and Achilles is at some particular spot, and neither is at the same place twice during the race. The tortoise will be in the same number of places that Achilles will be in, because each one is in a particular place at a particular instant and in another place at another instant. However, for Achilles to get ahead of the tortoise, the condition must be met that the places in which the tortoise will be, will be just a portion of the places where Achilles will be, because the tortoise began with an advantage."

Now focus and pay attention. Russell's version can be disproved only if you ignore the axiom which states that the part is always smaller than the whole: Achilles was in only *some* of the spots that the tortoise was in. Are you unwilling to give it up? Russell points out that whoever believes in this axiom must also agree that Achilles, even if he runs ten times, a thousand times, or even a million times faster than the tortoise, will never be able to catch up as long as the tortoise gets an advantage of a metre or a centimetre or a millimetre.

What's going on here? Are you following me? I can show you that there are infinite points on the route over which both the tortoise and Achilles pass. Perhaps when we are talking about the infinite, the rules that we are accustomed to lose their validity?

By the way, if you recall my first paradox, this whole argument is moot. Achilles and the tortoise can't even begin the race – motion is impossible. I was just being nice to you by allowing you to make these really strange assumptions. Hah! You couldn't even get them started. You couldn't even shoot the starting pistol. To be able to press the trigger, your finger would first have to move half the distance, and then half of what remained and again and again … the same reasoning.

Once I was late for a meeting with my great teacher and master Parmenides. I explained to him that I was late because I had to pass infinite half-distances to get to our meeting at the Fair Helena Tavern. We were both amazed that I had arrived and that we were conversing with each other.

Truthfully, I have no idea why I am even bothering to justify my explanations to you. As Chinese philosopher Lao Tzu once said: "The wise man doesn't argue, and he who argues is not wise." I am wise, so I am getting out of here (if I can).

Paradox 3: The arrow project – rest and motion

In Zeno's third paradox, he "proves" that since an instant cannot be divided, an arrow shot from a bow will be, at any instant, in a state of rest (because if the arrow was in motion at any given instant, this would be due to the fact that an instant *is* divisible).

If we now assume that time is made up of instants, and that at any particular instant the arrow is static, we will have to conclude that the arrow will never be in a state of motion and therefore (and here again, Zeno is going to amaze us and we must get ready to face slings and arrows from him) it can never cover any distance.

At any instant we choose, the arrow must be at rest. How then is it possible that all those rests add up to motion? If the arrow moves a distance equal to zero at every instant, how do all these zeroes add up to a positive number and "allow" the arrow to fly?

Things are not simple!

Until today, there is no solution to this paradox; that is to say, no solution that would be acceptable to all members of the communities of physicists and mathematicians.

The gallant gait of Ms Gadot

Let's examine another version of this paradox. Let us assume that Wonder Woman Gal Gadot is in Tel Aviv walking down Rothschild Boulevard. Nobody will be in the least surprised if I would say that hordes of people are following the beautiful lady and photographing her from every possible angle. All of a sudden, Instagram is loaded with hundreds of photos of her, and in all of them, this gorgeous woman is in a static position, which is to say in a state of rest. That is the nature of a photograph: it grabs a specific instant and perpetuates it for eternity. If there is something moving in the pictures, you had better replace your old camera for a newer model or read the manual to find the setting to speed up the shutter.

Since one can photograph Gal at every single instant, it stems from this that she was in a state of rest for her entire walk down the boulevard. So we must ask: "If she is always in a state of rest, when exactly did she walk? How do all these rests end up being motion?" Again, the same question. And again, the answer is not that clear.

Dealing with Zeno's paradoxes

CHILDHOOD MEMORIES – ZENO IN A GEOMETRY LESSON (A PRE-POST-SOCRATIC DIALOGUE)

Zelia the Teacher: As you remember, children, only one unique straight line passes through any two points.

Zeno: No line passes through any two points, because it is impossible to go from any point to any point. I've already clarified this a number of times. I also don't understand why you rejected my brilliant solution to the problem about the ship that sails from Magra to Athens – despite the short distance, it will take the ship an infinite amount of time to reach its destination. That is to say, it won't. You simply are unable to think outside the standard curriculum.

Zelia: Zeno, you constantly argue about the simplest and most obvious things and complicate everything needlessly.

Zeno: Nothing is simple and clear.

Zelia: To what are you referring this time?

Zeno: Last lesson, you taught us that a line is made up of infinitely many points, true?

Zelia: Very true.

Zeno: Did you not also say that the length of a point is equal to zero?

Zelia: Of course. Because if it was anything else, the point would be able to be divided, which contradicts our basic premise. If it has any length, it is a line, not a point. Also, it is impossible that a point has any length due to the fact that between any two points there is always another point – in fact additional points. If a point had a length that is greater than zero, and two points were separated from each other by less than this length, it would be impossible to place the first point between the other two. Which totally contradicts all basic logic of geometry.

Zeno: Fine. You don't have to try so hard. I agree with you that the length of a point is equal to zero. Now I want to ask a little question: How is it possible that a line of any length, say 17 cm, is made up of points whose lengths are zero? Already in grade 1 we learned that the sum of any number of zeroes is always zero. So I repeat my question: How is it possible that a set of points that are zero in length can form a line that is 17 cm long? I am waiting for clarifications and answers.

Zelia: I need to think about that a bit. I'll give you my answer next lesson.

Zeno: Don't hurry. I can wait. Here is another similar problem that may help you in your quest. A square consists of an infinite quantity of lines, each with an area of zero. How is it possible that these lines can manage to fill a square with a positive area? Maybe you should go and discuss it with Zelotes, the physics teacher. Ask him in the language he understands: "How is it possible that an arrow that in time $t = 0$ covers a distance of $s = 0$ can travel from one place to another? Is it not true that at any given instant it is travelling a distance of 0? One can photograph the arrow – yes, I know, photography hasn't been invented yet – and note that at any particular instant, it is at rest. Perhaps time is not comprised of instants? Perhaps if there are enough zeroes, their sum does not necessarily add up to zero?" Anyway, I am going out to test my theory using my slingshot.

(The bell rings. All the pupils flee with great delight from the classroom to the schoolyard. The teacher quickly escapes to the staff-room to sip some ouzo and unwind. Only Zeno remains in the classroom, thinking about arrows, about slingshots, and about tortoises that are faster than famous, fleet-footed heroes. Anyway, it is clear to him that he cannot really leave the classroom, since he would first have to go half the distance...).

Henri-Louise Bergson (1859–1941) was a Jewish French philosopher who had a great influence on philosophical thinking in the first half of the twentieth century. He was convinced that the human mind was unable to come to terms with Zeno's paradoxes and would never be able to. Bergson believed that the only thing to do was to formulate a practical approach to deal with them. However, there were other Frenchmen who had much less patience for Zeno and his paradoxes. For example, Henri Poincaré (1854–1912), an important mathematician, theoretical physicist, and scientific philosopher, had this to say:

> *Zeno was an idiot, and only idiots can deal with his paradoxes.*
>
> (Henri Poincaré)

The Englishman Bertrand Russell doesn't agree with the Frenchmen; in his *The Principles of Mathematics*, Russell wrote that Zeno's paradoxes are "immeasurably subtle and profound".

THE POINCARÉ CONJECTURE AND THE PERELMAN REFUSAL

The Poincaré conjecture is one of the seven open mathematical problems of the millennium declared by the Clay Mathematics Institute in 2000. At the time of writing, it is the only one of the seven that has been solved.

The problem was solved by the brilliant Jewish Russian mathematician, Grigori Perelman (born 1966). For his achievement in solving the Poincaré conjecture, Perelman was to have been awarded the Fields Medal and the Clay Millennium prize of one million dollars. I say "was to have been" because Perelman declined both prizes. "I'm not interested in money or fame. The accuracy of the proof is the only thing that is important," explained Perelman.[1]

He never even published a paper with his proof. Other mathematicians have written about it.

Perelman is a notorious refuser of prizes and rewards. He also refused a prestigious prize from the European Mathematical Society, claiming that those who were awarding the prize were not able to understand and appreciate his work.

In 2003, when he was only 37 years old, Perelman retired from mathematical research. Today, he is unemployed and lives with his mother in Saint Petersburg.

From Zeno to Newton, from Galois to Perelman, great mathematicians, for the most part, are quite unlike regular human beings. Perhaps this is part of what makes them great mathematicians.

The time has come to return to our topic of conversation.

Story time – Achilles gets disqualified

After Achilles failed in his attempt to overtake the tortoise, he decided to begin an intensive training program. Achilles came to the Olympic stadium and mapped out his running course, which would begin at point A and end at point B. However, an infinite number of gods decided to interfere with Achilles in the completion of his task. The first god decided not to allow Achilles to cover half of the distance, the second god decided not to allow Achilles to cover a quarter of the distance, the third god … the eighth … and so forth.

Brain-twister

Assuming these gods can do anything they want, prove that Achilles cannot even begin to run the course toward point B.

Brain-twister, continued

If you have concluded that Achilles is even unable to move, what is the reason for this? As long as Achilles is standing at the starting line, none of the gods are interfering. So what or who is stopping the (almost) undefeatable warrior?[2]

A staccato race

Imagine that a tiny little change is introduced into the race between Achilles and the tortoise. Each time Achilles reaches the point where the tortoise had been before, the competitors will stop and rest a minute (the tortoise really needs this). In this case, Achilles will catch up to the tortoise in infinite minutes, that is to say, he will never catch up. So many variations on a theme!

The death of a hero

The day after the staccato race, an extremely disheartened Achilles decided that, at any rate, he would continue his training. He decided to start training at two in the afternoon. A minute before two, Achilles reached the stadium. However, as usual, nothing is simple with our hero from the Trojan War. The infinite gods up on Olympus were furious over the racket the warrior was making just when they were getting ready for their sacred afternoon siesta, and they decided to do away with him by shooting him in the heel with a poisoned arrow. The first god decided to shoot Achilles a half a minute *after* the stroke of two, the second god decided to kill him at a different time – a quarter of a minute after two o'clock. The third god chose to do the dastardly deed at one eighth of a minute after two … and so on.

At two o'clock and one minute Achilles's state of health is in a very bad way – he was lying dead with infinite poisoned arrows embedded in his heel. However, none of the gods could be blamed for his death. Each of them had an excellent and identical excuse: "When I shot my arrow at Achilles, he was already dead with infinite arrows embedded in his heel. I admit that shooting a corpse is not a nice thing to do, but it is a long way between this and accusing me of murder."

Question: Who murdered Achilles? When?

Mathematicians in space

You have probably already noticed that as soon as we begin to touch on concepts like zero and infinity, many of the "normal" laws do not apply anymore. I will tell you about a famous thought experiment that is called "The spaceship".

Try to imagine what would happen to a spaceship that travels according to the following rule: In the first half hour, it flies at a speed of 2 km per hour (really quite slow for a spaceship). A quarter of an hour later, it goes at a slightly higher speed – 4 km per hour. For the next one eighth of an hour, it flies at 8 km per hour, and so forth. Where will the spaceship be after one hour?

The calculation isn't complicated. In the first half hour it is flying 2 km/hour and will go one kilometre. In the next quarter of an hour, flying at a rate of 4 km per hour, it will go 1 km. And so on and so forth: another one kilometre and another one kilometre and another one. It is easy to see that the distance the spaceship travels is equal to 1 + 1 + 1 … However, we need to add infinite ones, and therefore the result is infinity. So where is the spaceship? It seems that it is nowhere, because the spaceship is supposed to be at an infinite distance from its blast-off point. If the spaceship is at any one particular point, it is at a particular point from the blast-off, but this cannot be because it has gone an infinite distance. So where is it? Nobody knows. The search for the spaceship continues to this day.

Infinity cannot be found anywhere on the infinite line.
(George Hegel)

Verbal calculations
Prove without actually calculating anything (i.e. verbally) that:

$$\frac{1}{3}+\frac{2}{9}+\frac{4}{27}+\frac{8}{81}+\frac{16}{243}+...+...=1$$

THE BIOGRAPHY OF TRISTRAM SHANDY

(DEDICATED TO THE MEMORY OF ACHILLES)

One of the craziest, most bizarre stories I ever read is the story by eighteenth-century, Irish-born English writer Laurence Sterne, *The Life and Opinions of Tristram Shandy, Gentleman*, published in nine volumes in 1759–1766.

As you may guess from the name, the book's hero decided to tell us his life story and give us a treatise on his thoughts and opinions. But it is not exactly what you think.

The book has nine volumes, and one of its absurd ideas is that no event of the protagonist's life is simply told – every little thing that is even remotely connected with an event must be told. It is only in the third volume (!) that we get to the moment of his birth.

Tristram complains that he needs a year just to tell us about one day in his life. Bertrand Russell once pointed out that if the gentleman had lived infinitely many days, he would not have had a problem relating the entire story of his life. Really? On the one hand, every day in his life would finally get its turn to be told. He would tell about the 10,000th day of his life in his 10,000th year. On the other hand, every day that passes increases the gap between his real life and his written life by another year. What we have here is a type of Achilles-and-the-tortoise race, where Achilles runs a year while the tortoise just a day, but because of the infinite time at his disposal, the tortoise will eventually reach every place that Achilles did.

INFINITY AND BEYOND: THE NEVER-ENDING JOURNEY

> From time immemorial, the infinite has stirred men's emotions more than any other issue. Apparently, no other idea exists that inspires yet violates the human mind more than the concept of infinity, yet it is precisely for this reason that it needs clarification more than any other concept.

These are the words of the great twentieth-century mathematician David Hilbert (1862–1943) in his paper "On the Infinite".

If you carefully examine the first part of the book you are now holding in your hands, you will observe that "infinity" is indeed its shining star – the kingdom of numbers is an infinite kingdom, and many – perhaps even most – of its riddles and secrets are connected, directly or indirectly, to the concept of infinity.

Mathematics is the science of infinity.

(Hermann Weyl)

LESSON 6

GEORG CANTOR'S KINGDOM OF INFINITY – SET THEORY

Love from the third lesson

The subject that fascinated me most in my first year of mathematics studies in university was "set theory". The name is not particularly inviting and does not even begin to hint at the topics studied in this course. The outset did not bode well: amazingly boring definitions, axioms and relations. However, after the first two lessons, I realized that the course should actually be called "infinite set theory", since it discussed the infinite, in defiant disregard of Galileo Galilee's admonition not to do so. In addition, the discussion on infinity was not cloaked by any metaphysical or theological aspect – I was already familiar with such perspectives, and some of them, such as Kant's antinomies, or the concepts of Nicholas of Cusa, or the world vision of Giordano Bruno (who was greatly influenced by Nicholas of Cusa), had quite inspired me. But now it was all very different, and I could sense that something unlike anything I had known before was unfolding. I sensed that something wonderful was about to be revealed.

The lessons on set theory and its star, infinity, which were given in a most fascinating and thought-provoking manner by my late teacher Mordechai Epstein (to whom I owe a huge debt of gratitude), expounded on the topic in a precise and purely mathematical manner. Suddenly I discovered that it was possible to compare different forms of infinity: there were actually larger infinities and smaller infinities and, in fact, there is an *infinite* range of infinities! I was spellbound.

What kind of amazing person is so intimate with infinity that he can recognize and distinguish between its various forms? Well, this individual was Georg Cantor, and set theory, which he developed, is often called "Cantorian set theory" in his honour.

Georg Cantor – The man who saw infinity

Georg Ferdinand Ludwig Philipp Cantor was born in 1845 in Saint Petersburg. He began his academic career in 1862 at the University of Zurich. A year later, after the death of his father had left him a not insubstantial inheritance, Cantor transferred to the University of Berlin to study mathematics, physics and philosophy. Cantor spent the summer of 1866 at the University of Göttingen, the most important centre of mathematics at the time. (The university retained this title until the Second World War.) In 1867, Cantor received his PhD from the University of Berlin for his work on number theory. Cantor taught briefly at a girls' school in the city, then later went to work at the University at Halle, where he remained until the autumn of his life. In 1872, Cantor met Richard Dedekind, a meeting that led to both a personal and professional friendship.

In 1874, two major events occurred in Cantor's life. The first was his marriage that would bring six children into the world. The second was the publication of his revolutionary paper on the topic of infinite sets, "Über eine Eigenschaft des Inbegriffes aller reellen algebraischen Zahlen" ("About typical properties of real algebraic numbers"). The title is not very "user-friendly" and I imagine that my translation of the title into English doesn't add much to your understanding of what the paper was about. Nevertheless, there is no argument that this paper launched the study of set theory and, for 25 years, served as the discipline's cornerstone.

This paper was the first to suggest that there possibly might be more than one type of infinity. It saw publication despite sharp criticism from Leopold Kronecker (1823–1891), Cantor's former teacher and a very influential mathematician at the time. Kronecker disapproved of Cantor both on professional and personal levels, going so far as to insult him by calling him a "mathematical charlatan" and "a destroyer of youth". (It is impossible not to recall that Socrates was also thus labeled. It seems a good sign.)

I don't know what predominates in Cantor's theory – philosophy or theology, but I am sure that there is no mathematics there.

(Leopold Kronecker)

I have no idea why Kronecker saw fit to *accuse* Cantor of introducing theology into the discussion of mathematics, considering that one of Kronecker's most famous statements refers to those entities residing above:

God made the integers, all the rest is the work of man.

(Leopold Kronecker)

Kronecker was not Cantor's only critic. Poincaré, the French mathematician who, if you remember, ridiculed Zeno's paradoxes and anyone who involved themselves with them, also emphatically opposed Cantor's thinking. Poincaré believed that Cantor's ideas were a disease that contaminated the legitimate discipline of mathematics.

Swedish mathematician Magnus Gustaf Mittag-Leffler (1846–1927) actually appreciated Cantor's ideas, but he thought they were ahead of their time and should be published a century later. To this Cantor responded that in his opinion, waiting a hundred years was "too much of a demand". Cantor was very sensitive to criticism of him and his theory. In 1884, he suffered his first bout of intense depression.

There is an urban legend that maintains that Mittag-Leffler is the guilty party behind the fact that there is no Nobel prize for mathematics. Various theories are presented: Nobel was in love with Mittag-Leffler's wife, or Nobel's mistress betrayed him with Mittag-Leffler, or Alfred Nobel despised Magnus Gustaf for personal reasons.

Cantor's method of coping with this emotional crisis was quite unique and seems appropriate for a man as special as him: he decided to temporarily leave mathematical research and focus instead on Elizabethan literature. To this he dedicated much time and effort, as he attempted to substantiate the thesis that Francis Bacon wrote the works attributed to William Shakespeare. In 1896 and 1897, Cantor actually wrote two papers on this subject.

In nature's infinite book of secrecy
A little I can read.

(William Shakespeare)

Cantor's involvement with Shakespeare helped him recover, and he returned to his true calling – the subject of infinity. In 1891, he published a paper presenting an elegant and brilliant concept that is today called "Cantor's diagonal argument". (We shall examine it shortly.)

However, Cantor's emotional situation did not give him peace. In 1899, he was admitted to hospital. In that same year, his youngest son died suddenly, his depression became chronic, and Cantor's interest in mathematics and infinity almost completely waned. In 1903, he was admitted to hospital once again.

A year later, an event occurred that, according to Cantor's biographer, history professor Joseph W Dauben, shocked Cantor so much that it made him doubt the existence of God. This is the place to mention that Cantor believed that God had bestowed upon him the theory of infinity, and Cantor's task was to pass it on to ordinary mortals. What follows is a summary of the incident.[1]

In the late 1890s, Georg Cantor and German mathematician Felix Klein took it upon themselves to found the International Congress of Mathematicians (ICM). Klein even penned a slogan in the style of Karl Marx: *Mathematicians of the World, Unite!* Right up until today, these congresses are the major mathematical events in the world; the prestigious Fields Medal and the Gauss Prize are awarded at the opening ceremony.

The first ICM took place in Zurich in 1897. The second took place in Paris in 1900 and is famous for the 23 open problems presented by David Hilbert. (The first problem on the list concerned the continuum hypothesis, the problem that Cantor had presented back in 1878, and which we will discuss momentarily.)

We are now at the third congress, which took place in Heidelberg in 1904. Cantor and his daughters are sitting in the audience. On stage comes Hungarian mathematician Gyula Kőnig, who pronounces that Cantor's theory suffers from basic errors. Cantor is deeply shocked to have been humiliated in front of his daughters and colleagues. In fact, Kőnig had ignored the most basic rule of mathematics: precision, and the very next day, mathematician Ernst Zermelo,[2] one of the founding fathers of set theory, asserted that Kőnig was wrong and talking nonsense. But this did nothing to alleviate Cantor's feelings.

Cantor retired from the university in 1913 and suffered abject poverty during the First World War. He died in 1918 in a sanatorium in Halle.

At the outset of the twentieth century, there were still sharp disagreements regarding the importance of Cantor's theory and its accuracy. Nevertheless, in 1904, Cantor was awarded the Sylvester Medal, the highest award given by the Royal Society to a mathematician and named after the English mathematician James Joseph Sylvester. Ironically, the previous recipient of this award was Cantor's bitter rival, Henri Poincaré.

Mathematics is the music of logic.

(James Joseph Sylvester)

Among Cantor's most ardent admirers were Bertrand Russell (1872–1970)[3] and David Hilbert, who called

Cantor's set theory "the greatest product of mathematical genius and human thinking".

> *No one shall expel us from the paradise which Cantor has created for us.*
>
> (David Hilbert)

> *Cantor's Paradise is a fool's Paradise.*
> *His theory is ridiculous and total nonsense.*
>
> (Ludwig Wittgenstein)

(Clearly, sometimes even great philosophers speak nonsense.)

Cantor's apology

> *My theory stands as firm as a rock; every arrow directed against it will return quickly to its archer. How do I know this? Because I have studied it from all sides for many years; because I have examined all objections which have ever been made against the infinite numbers; and above all because I have followed its roots, so to speak, to the first infallible cause of all created things.*
>
> (Georg Cantor)

Today, the importance of Cantor's set theory is obvious to nearly everyone involved in higher mathematics. Modern versions of set theory that have been evolved as a result of his pioneering studies are today used as the foundation for a considerable number of mathematical theories developed in the twentieth century.

It's time to meet Georg Ferdinand Ludwig Philipp Cantor's set theory.

An introduction to set theory: What is a set?

In this and the following sections, we will try to understand the central ideas of Cantorian set theory. Let us start with the most basic concept – the set. What is "a set"?

Here is the intuitive definition that served mathematicians at the dawn of "set theory" days:

> **Definition: Set**
> **Any collection of items.**

This definition seems a bit too generalized. It does not even pose a requirement for some common factor among the objects that make up the set. It is not surprising, therefore, that this definition led to not a few problems over the course of time.

How does one denote a set? One way is to list all the items in the set. For example, A = {Gustav Mahler, Gustav Klimt, Gustav Eiffel, Gustav Holst, Gustavo Dudamel, Gustave Doré, Gustavo Boccoli, Gustave Courbet, Hurricane Gustav, Gustaf V of Sweden}. This set has exactly ten members, and all its members have one thing in common – the word "Gustave" or some form of it.

But a common factor is not a necessity. Here is another example of a perfectly good set: B = {1729, a, 4, {4}, Pushkin, Pushkash, $, set} (It is simply a set of eight seemingly random items, as listed.)

It is important to be able to discern whether an item is a member of a particular set or not. The Swedish mathematician Magnus Gustaf Mittag-Leffler is *not* a member of set A, even though his name includes a form of Gustave, because he was not defined as a member of the set, but the dollar sign is a member of set B, because it *is* listed as part of that set.

This method – that is, listing all the members – will not be a very satisfactory method for defining the set, say, of all the even numbers. Therefore, another way to define a set is the use of ellipses. This is the way we could denote the set of all even numbers: E = {2, 4, 6, 8…}. However, the "rule" that is indicated by the ellipsis is not always apparent and unambiguous to all. For example, look at this set: T = {1, 3, 6, 10, 15…}.This is the set of triangular numbers. (The letter designation for the set provides an additional hint.) However, this may not be obvious to everyone. Although even those not familiar with the concept of triangular numbers might guess how to continue the series.

But this is not always so. Here is another example. F = {1, 3, 9, 33, 153…}. What values should appear in place of the ellipsis? Have you guessed?

Here is the answer:

$1! = 1$
$1! + 2! = 3$
$1! + 2! + 3! = 9$
$1! + 2! + 3! + 4! = 33$
$1! + 2! + 3! + 4! + 5! = 153$
Therefore the next number will be
$1! + 2! + 3! + 4! + 5! + 6! = 873$
and so on.

Another way to define a set would be to point out the common property by which the items in a set are defined. For example, "the set of all the past and present NBA players", "the set of all the atoms in the universe", "the set of prime numbers", "the set of happy people", "the set of all even numbers that cannot be presented as the sum of two prime numbers", "the set of numbers that are greater than themselves", "the set of Sumo wrestlers that weigh more than 250 kilograms", "the set of all the films directed by Andrei Tarkovsky", "the set of all poems written by Arseny Tarkovsky" (the poet Arseny Tarkovsky is the father of the great Russian film director Andrei Tarkovsky), "the set of all numbers with interesting properties" and so forth.

As you most probably have already understood, it is customary to denote a set using capital Latin letters A, B, C, D …

The symbol ∈ denotes membership in a set. For example, if we denote by "F" the set of all films directed by Fellini, we could write *Amarcord* ∈ F. The same symbol struck through denotes an item that is not part of the set. For example, *Avatar* ∉ F.

In Cantor's set theory, every item will or will not be a member of a particular set. But think, for a moment, about the set of, say, tall people – here it is not very easy to determine membership in the set. In 1965, American-Jewish mathematician and computer scientist Lotfi A Zadeh (1921–2017) came up with a more flexible approach for sets which is called "fuzzy set theory". The basic concept of fuzzy set theory is that any item can be assigned a specific probability between 0 (for sure not) and 1 (absolutely certain) of being a member of any particular set. The classic approach uses only the numbers 0 and 1. For example, Napoleon and Danny DeVito have 0 probability of being part of the set of tall people, and LeBron James has a 1 probability of being part of that set, whereas the author of this book has about a 0.07 probability of being included in the set of tall people.

My introduction to fuzzy set theory came when I happened to read Bart Kosko's book *Fuzzy Thinking*. I loved the book already from its very first line: "One day I learned that science was not true." The main premise of the book, which the author defends brilliantly throughout its 300 pages, is that there is nothing in our world that is black and white. Rather, all things are in various shades of gray. Only in classic mathematics are things absolutely certain, but classic mathematics cannot faithfully describe the world.

Here is a quote by someone who summed up the idea much better than I can:

> *As far as the laws of mathematics refer to reality, they are not certain; and as far as they are certain, they do not refer to reality.*
> *Insofar as mathematics is about reality, it is not certain, and insofar as it is certain, it is not about reality.*
>
> (Albert Einstein)

But let us return to Cantorian set theory.

How many elements are there in the set of all the even numbers that cannot be written as the sum of two prime numbers? Hopefully, you recall Goldbach's conjecture, which claims that there are no such numbers. In other words, there are no items in this set. A set that does not have even one element is called an "empty set" and is denoted thus: $\emptyset$.

One can say that the set of all even numbers that cannot be written as the sum of two prime numbers has a very high probability of being an empty set, yet there is no absolute certainty that this is indeed the case. On the other hand, the set of numbers that are larger than themselves and the set of hedgehogs that speak Yiddish are most certainly empty.

Okay then, perhaps it is possible, similar to fuzzy set theory, to exhibit some flexibility with respect to whether or not an element is a member of a particular set, but up until now we have not seen anything especially out of the ordinary that would prevent us from defining a set as some collection of elements. Furthermore, I will point out that it was Cantor himself who gave a similar definition to the concept of the set.

However, almost nothing is really simple and self-evident in the world of mathematics – it only sometimes seems so at first glance.

"Obvious" is the most dangerous word in mathematics.

(E T Bell)

It turns out that this intuitive definition of a set as some collection of elements has a number of traps. As an example, I will offer the bitter experience of German mathematician, logician, and philosopher Gottlob Frege (1848–1932).

In 1902, Frege was about to publish the second volume of his monumental work *Grundgesetze der Arithmetik* (The Basic Laws of Mathematics), in which he showed how regular arithmetic can be reconstructed from Cantor's foundations of set theory using only the naive definition Cantor used to define a set. Frege sensed that all his work was about to fall by the wayside when, on 16 June, he received a letter from Bertrand Russell, in which Russell offered a paradox that he had formulated and which has since become very famous. It is known as "Russell's paradox" or "Russell's antinomy".

To shave or not to shave? – Russell's paradox

There are many versions of Russell's paradox, but the most famous one is known as the "Barber's paradox".

In a small, remote English village lives Edward, a barber by profession, who is known for being extremely pedantic. Years earlier, when he opened his barbershop "Edward Scissorhands", he proclaimed a rule: he would shave *all* those people in the village who did not shave themselves, and only them.

All was well and good the first day. There were those who shaved themselves, and there were the others, who came to Edward to get the kind of close smooth shave only his skilled hands could give. On the second day, Edward began to notice a layer of stubble on his cheeks and chin that was not very attractive. However, an instant before picking up the razor, the punctilious barber realized he had a problem due to the rule that he had himself imposed.

According to the rule, he was only supposed to shave someone of the village who did not shave himself. So was he allowed to shave himself? To shave or not to shave? That was the question.

Pay attention to what is happening here. If he shaves himself, then he is breaking his rule, because he will be shaving someone who shaves himself; but if he doesn't shave himself, he is a member of the village who doesn't shave himself, and such a person must be shaved by him.

Russell's paradox is the result of a principle known as a "vicious circle". Based on this principle (assuming you want to avoid such a paradoxical incident), it is not a good thing for a set to include an element that can be described using the definition of the set itself.

A fascinating analysis of this paradox is given in Raymond Smullyan's book *Alice in Puzzle-land: A Carrollian Tale for Children under Eighty* (Humpty Dumpty explains the paradox to Alice). Smullyan's conclusion: The barber paradox is equivalent to the claim "I know someone who is short and also tall".

Here is another version of Russell's paradox. A librarian decides to compile two catalogues for her library: one is yellow and is called "The Yellow Catalogue of Books that Mention Themselves" and the other is "The Blue Catalogue of Books that Don't Mention Themselves". One by one, the librarian examines each book in the library and enters its title into either the yellow or blue catalogue. The latter is very large and the former quite thin, since most books don't mention themselves. Now the librarian is down to the final two books to be catalogued: those very same yellow and blue catalogues. The Yellow Catalogue can be entered into either itself (because that means it is self-referencing, and that's fine), or into the Blue Catalogue (since then the Yellow Catalogue is not self-referencing, and that's also fine). But what to do with the Blue Catalogue, the one that lists all books that *do not* make a reference to themselves? If it is listed in itself, this means that the Blue Catalogue is referencing itself, and it shouldn't be in there. However, if it is catalogued in the Yellow Catalogue, the Blue Catalogue has not listed itself … so it shouldn't be included in the Yellow Catalogue, which is for books that list themselves. Clearly, we have reached an impasse. No matter what we do with the Blue Catalogue, we violate the rule.

I don't want to belong to any club that will accept people like me as a member.

(Groucho Marx)

Two types of sets

Let us return to the issue at hand. There are two types of sets. The first type are called "standard sets" and these are sets that don't include themselves as an element. The set of all rabbits is an example of a set of this type, because the *set* of all rabbits is not a rabbit and therefore will not be an element in itself.

On the other hand, the set of all "non-rabbits" is a set of the second type, which is sets that include themselves. A set of "non-rabbits" is itself not a rabbit. Similarly, "sets of objects that can be described in exactly eleven words" is also a type of the second set. These sets have an unusual property in that they themselves meet the defined properties for their members. In simpler terms: sets of this second type include themselves as elements. Think, for example, of the set of all the ideas that one can imagine. This set includes itself as one of the elements – obviously, a set of all the ideas that can be thought about is also an idea. In honour of Russell, it is customary to symbolize this second type of set using the letter R. That is to say, any set that can include itself as an element is today called a "set of type R". Every set *must* be either a standard set or an R set, meaning that, allegedly, any particular set cannot be both a standard set and an R set at the same time.

But is this a fact?

Let us observe now the *set of all standard sets*. We shall call this set M. And here comes a surprise: set M is not a standard set, but it is also not an R set. I shall explain.

If M were to be a standard set, then it would have to be included as an element within the set of standard sets, namely M. But then M is an element of M, thus M cannot be standard because it is included in itself and is therefore an R set. We have reached a contradiction.

On the other hand, if M is a set of type R, this means that it does *not* belong to the "set of standard" sets. But that is exactly what M is! Again, we have a contradiction.

As you can see from all this, Cantor's original "intuitive" definition of a set using natural language is called "naive set theory" and can lead to paradoxes that cannot be resolved. Therefore, other methods of defining sets are used today.

We can conclude from all this the following:

1 Free use of an intuitive definition for the notion of a set can lead to unwanted paradoxes.
2 One should not set rules that the public cannot abide by.
3 The set of all "non-rabbits" is a set that is much too large to discuss.

TWO ASIDES: ONE VERY SHORT AND THE OTHER SLIGHTLY LONGER

1 Because of the great affection I have for Italy, I cannot resist mentioning that Italian mathematician Cesare Burali-Forti (1861–1931) had previously discovered something similar to Russell's paradox in 1897. He was

involved in research in set theory and was testing a concept called "the set of all the ordinals".

2 French philosopher Jean Buridan also presented a paradox in the fourteenth century that is very similar to Russell's barber paradox. In Chapter VIII of Sophismata, entitled "Insolubilia", Buridan offers the following anecdote:

Plato controlled his disciples on a bridge and did not allow anybody to cross without his permission. One day, Socrates arrived at the bridge, and demanded that Plato let him pass. Plato did not like the tone of his teacher's voice and said to him: "If the first statement you say is the truth, I shall let you pass; but if the first thing you say is a lie, I swear that I will throw you into the turbulent water." Socrates thought for a bit and said: "You will throw me into the water."

Let's examine what's going on here. If Plato throws Socrates into the water, then Socrates told the truth, and so Plato shouldn't have thrown him into the water; he should have let Socrates pass. On the other hand, if Plato lets Socrates pass safely over the bridge, then Socrates lied, and therefore Plato should have introduced him to the swirling waters of the river.

So much for Buridan.

By the way, an almost identical paradox appears in chapter 51 of the second volume of *Don Quixote*, when Sancho Panza is appointed as governor of the Island of Barataria. You might like to take a break from mathematics for a bit and read this wonderful chapter. Enjoy.

LESSON 7

HILBERT'S GRAND HOTEL *INFINITY*

On the planet Proxima Infiniti, which is at a distance of thousands of light-years from Earth and everywhere else, there is one of the wonders of post-modern architecture. Built there is a hotel based on the concepts of David Hilbert the mathematician. The hotel is named after him and is part of the luxurious Hilbert chain of hotels. Each floor of the hotel consists of one suite that houses one guest only. The hotel has an infinite number of rooms, yet despite infinite rooms, the hotel rises to a height of only one metre. The Zeno Ltd Company built the hotel thus: the height of the first floor is half a metre, the height of the second floor is a quarter of a metre, and the height of the third floor is an eighth of a metre, and so on and so forth – up until infinity. Try to imagine how this hotel looks from a bird's eye view.

The size of the rooms does not bother the hotel's regular patrons, who are the natural numbers: 1, 2, 3, 4, 5, 6 … Each number inhabits the floor that bears his number. The hotel is always at full capacity. The hotel staff, all of whom live in the lobby of the adjacent building, includes Omega, the reception manager; Epsilon, the reception clerk; and two chambermaids, the sisters Sigma and Lambda (Sigma is responsible for the odd-numbered rooms while Lambda is responsible for the even-numbered).

One day, 0 arrived and asked the reception clerk if there was a room available for him. Epsilon replied that he was very sorry but the hotel was totally full and there was absolutely nothing he could do. "To get a room at this hotel, you must reserve far in advance. It's not for nothing that we were awarded five stars."

Luckily for our guest, the reception manager arrived. Omega berated the reception clerk and explained to him that if the hotel were a finite hotel, then, of course, there would be nothing that could be done. But this was an infinite hotel, and the problem could be solved very simply.

Omega announced on the PA system: "Dear guests. Each of you is kindly requested to move to the floor above you." All the numbers complied with Omega's instructions. Number 1 moved to the second story, 2 moved to the third, 3 moved to the fourth story, and so on. The first story was now vacant, and number 0 could now count himself as a guest of the hotel.

"Please note what I did!" said the manager to the astounded Epsilon. "I trust that if any finite number of guests arrive in the future, you will be able to receive them yourself."

"Of course," answered the clerk. "If a thousand new guests show up at the hotel. I will simply ask each of our longtime guests to move to a room whose number is greater by a thousand than the one he is in at the moment, and thus the correct number of rooms will be vacated."

Satisfied, Omega returned to her other occupation, but her repose was interrupted by the ring of the telephone. On the other end of the line she heard the angry voice of 13.

"I want to complain about the outrageous prices," said 13. "1,000 CS (cosmic shekels) a night is highway robbery. Especially considering that even if you charged only half a CS, your income wouldn't change – no matter what, your earnings are infinite."

"I'm sorry, but I can't reduce the price that much," the manager replied. "Just the daily maintenance of each room costs me 20 CS a day, and what about the salaries of Epsilon, Sigma, and Lambda?"

"That's not a problem," continued 13. "Even if your expenses are 500 CS per room per day, and you pay your staff 70 CS a second, you could still charge each guest only half a CS a night and also pay yourself a billion CS a day for spending."

"How's that?" asked Omega.

"Very simple. Although your expenses are infinite, so is your income. You can set aside a billion CS from the income anytime you want, and your income will still be infinite, and that way you can balance your expenses. A rich friend of mine has a bill in the denomination of an infinite number of CSs. I once saw him buy a newspaper and pay with this bill. Do you know how much money he got as change? He got his original bill back! Because the newspaper cost 2 CS and infinity less 2 is still infinity. That's how it is everywhere. The rich really don't pay for anything. But let's get back to our problem. Even if you reduce the price of a room to one-thousandth of a CS, nothing will actually happen. Even one-trillionth will do the job."

"Hmm…" mumbled Omega. "What you say makes sense. It sounds interesting … I promise you I will think about it. By the way, if I were to have an infinite-CS bill, I would retire immediately. Let's face it, no matter how much I earn, it wouldn't add anything to the fortune of an infinite CS."

Omega's daydreams about possible retirement were disturbed by a message that appeared on her computer screen. It went like this:

> To the Hilbert Hotel, from Alpha Negative.
> All of our residents, negative integers –1, –2, –3, …
> want to come for a one-week visit. We would
> appreciate it if you would find us room in your hotel.
> Sincerely yours, your faithful friend, –17.

Now the number of expected new guests was to be infinite, so the previous solution of moving residents by a finite amount was not feasible. You couldn't ask people to move an infinite number of stories up! It would take an infinite amount of time and nobody would know how far up to go and when to stop climbing.

"Perhaps we can skip infinite stories in a finite space of time?" Omega asked herself. "Let's assume that the first story can be passed in half a minute, the second story in a quarter of a minute, the third in an eighth of a minute, and so forth. Here! I have discovered a way to move infinite stories in only one minute. But where will the numbers move to? To which rooms? No. This will not do. I will have to think of something else."

The reception manager tried to think of another solution, but wasn't very successful. In the end, she decided to discuss it with the reception clerk, Epsilon. Perhaps if they put their heads together they could come up with an idea. However, this also didn't help.

Having no alternative, Omega decided to ask Lambda and Sigma to help. They had both taken advanced courses in algebraic topology and functional analysis before arriving at their present job.

"It's no problem whatsoever to accommodate them," said the sisters. "When we studied set theory, this was the first exercise we got. Here is the solution. The numbers 0 and 1 will stay where they are, on the first and second stories. All the other numbers will move to a room that is double their number. That is to say, 2 will go to room 4, 3 will go to room 6, 4 will have the pleasure of staying in room 8, and so forth and so on. Now all the odd-numbered rooms (3, 5, 7, 9 …) will be empty and we will have infinitely many vacant rooms. We will be able to accommodate all the guests."

Omega loved the clever sisters' suggestion so much that she put a sign in the lobby that read:

No vacancies.
This hotel is fully booked.

Rooms available.
We always have room.

And thus it passed. When the negative numbers arrived for their visit, everything worked like a charm. There was absolutely no problem accommodating them, and the hotel looked like this:

Room	1	2	3	4	5	6	7	8	9	…
Guest	0	1	−1	2	−2	3	−3	4	−4	…

I will explain the new room arrangement. 0 was in room 1, where he had been before the negative numbers arrived. Every other positive integer was in a room that was double its value. 3, for example, was in room 6, and 111 resided in room 222.

All the negative numbers were in rooms whose number was equal to the guest's number multiplied by (–2) plus 1. So –1 was in room 3, and –17 was in (–17) × (–2) + 1 = 35.

A peaceful week passed for the hotel and its guests.

When the negative numbers checked out of the hotel, 0 decided to leave with them. After they had all left, Omega was surprised to discover that the natural numbers, who had once filled the hotel to capacity, now only half-filled it, and that only the even-numbered rooms were occupied. In fact, she could cut expenses now and fire Sigma, who was responsible for the odd-numbered rooms, which were now totally empty. On the other hand, Sigma had helped her solve her problem of how to accommodate the negative numbers, not to mention that Omega didn't feel right about splitting up the sisters. However, the fact was that, one way or another, the occupancy of the hotel had gone down to 50 per cent, even though it had the exact same number of guests that had once filled it completely!

"Strange things are happening here," said Omega to herself. "And what would happen", she wondered silently, worry creeping up on her, "if number 1 went to stay in room 10, and 2 went to stay in room 20, and 3 in 30 and so forth? Our occupancy would be down now to only 10 per cent, even though not one single permanent resident had left the hotel! All the natural numbers are still here, but the occupancy rate would be so low that I would end up being fired."

While she was still contemplating this awful idea, she reminded herself that in another two weeks an important conference on "Positive Rationalism in the Age of Rational Positivism" was to take place in the hotel, and all the participants – all of the positive rational numbers – were supposed to be arriving for three days of a positive rational experience.

"There should be no problem finding space for them all," said Omega to herself. "The hotel is half-empty and there are infinite empty rooms.

Omega's serenity did not last long. She was suddenly overcome by a host of worrisome thoughts. Omega realized that the rational numbers with a denominator of 2 could fully occupy the hotel, with the fraction 1/2 to room 1, 2/2 to room 2, 3/2 to room 3, and so forth. But at the same time, the rational numbers with denominators of 3 or 4 or any other number could fill up the hotel in exactly the same fashion: 1/3 in room 1, 2/3 in room 2, 3/3 in room 3 … In other words, an infinite number of infinite sets were arriving, and any one of them could totally fill up the hotel. Not to mention that even before they arrived, there was already infinite natural numbers occupying the hotel (1, 2, 3, 4 …).

The manager tried to examine other possibilities, such as housing 1 in room 1, 2 in room 1001, 3 in 2001, … and then 1/2 in room 2, 2/2 in room 1002, 3/2 in 2002, and then 1/3 in room 3, 2/3 in 1003, 3/3 in 2003, etc … but it didn't take her long to realize that this plan also wasn't a good one. (Explain why this plan was doomed to failure.)

Since Omega didn't feel comfortable bothering Lambda and Sigma again – they were being paid for cleaning the rooms and not for being strategic advisers – she decided (she didn't have much choice) to ask for help from the head mathematician of the APEIRON system (where Proxima Infiniti is), Professor Finkelstein-Ostrovsky-Cantorovich.

The old but energetic professor said that the problem just looked complicated, but in fact was not.

"Thanks to a man named Euclid, who once lived on a small, blue, very distant planet named Earth, the problem can be solved relatively easily," said the professor with the triple name.

"What did Euclid do?" queried the reception manager.

"He proved that there are infinite prime numbers," answered the professor.

"And how does that help us find places for all our guests?" asked Omega, clearly doubting any possible connection between the infinite nature of prime numbers and finding a housing solution for all the guests of the convention.

"I will explain it to you as simply as I can," promised Finkelstein-Ostrovsky-Cantorovich. "The fact that there are infinite prime numbers allows us to solve, quite simply, the problem of accommodating all the rational numbers. Here is the plan. We will assign them their rooms according to prime numbers raised to successive powers as follows:

"The first prime number is 2. We shall send number 1 to room 2, and number 2 to room 2^2, and 3 will go to room 2^3, 4 will be housed in form 2^4 … and so on."

The professor continued: "The next prime number is 3. So the fraction 1/2 will go to room 3, 2/2 to room 3^2, 3/2 to room 3^3, 4/2 to room 3^4, and so on and so forth."

"But 2/2 is equal to 1, who is already in room number 1," contemplated the reception manager.

"This will be no problem at all. In fact, quite the opposite. Many rooms will be allocated to 1, and he will choose in which one of them to stay," replied the professor.

"Now we get to the third prime number, which is 5. So 1/3 will be given room 5." It was at this moment that the manager understood why the prime numbers must be raised to powers, because room number 4 was already occupied by number 2.

The professor carried on: "2/3 will go to room 5^2, 3/3 will be assigned 5^3 … you get my drift. Now we are at 7, the fourth prime number. It's the same idea. 1/4 goes to room 7, 2/4 to room 7^2, 3/4 will be able to check out the view from 7^3, and so forth, and so forth, and again so forth, and so on."

"This is a very interesting arrangement," the professor continued. "Even though there are infinitely many infinite groups of guests, each group of which could fill up the entire hotel on its own, we have succeeded in accommodating them all. And … and we still have an infinite number of empty rooms!"

"What!?" The reception manager could not believe her ears.

"Any room that is not a prime number nor a prime number raised to an exponent – for example 1, 6, 10, 12, 14, 15, 18 … are totally empty."

The manager, who only a moment before was ecstatic about the brilliant way that the professor had solved her accommodation problem, was now totally dejected. Once again, the per-cent occupancy problem had emerged. It is impossible that a good hotel manager should have infinite (!) empty rooms. What will the owners of the hotel say?

"Look," Omega said to the professor, "the natural numbers alone can fill up the hotel entirely by themselves, and once upon a time they did just that. Suddenly, you come along and suggest some crazy scheme by which all the natural numbers along with infinitely many infinite sets, each of which could also occupy the entire hotel on their own, leave me with much less than 100 per cent occupancy. This doesn't sound logical to me at all. I'm no expert, but is there perhaps some way to do something that can help me report to my superiors a much more favourable per-cent occupancy rate?"

"Oh, I thought that the solution would be much more impressive if infinite rooms remained empty, but if the only thing that interests you is occupancy rate, I can suggest a solution that will give you 100 per cent occupancy."

"Oh, please! Tell me what it is," Omega begged.

"Before I explain the solution, we will need to do some preparation. We will assign each rational number a pair of numbers. The first will be its numerator, and the second its denominator. For example, 3/4 is assigned the numerical pair (3, 4). We will consider the natural numbers, n, to be written as $n/1$, and thus they will be assigned the pair $(n, 1)$. For example, 7 will get (7, 1). Now, we arrange all the numbers as follows:

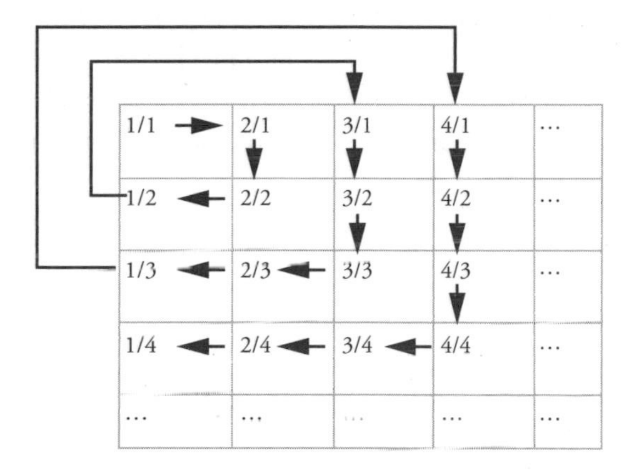

For lovers of algebra, I shall point out that in general, we assign number n/m to room number $n^2 - m + 1$ if $n \geq m$, or to room number $(m - 1)^2 + n$ if $n < m$.

"For example, 3/2 has a numerator that is greater than the denominator, and therefore he will be given room

$3^2 - 2 + 1$, that is to say, room 8. You can check that if we start out with pair (1, 1) and proceed according to the arrows (see the illustration above), the one with pair (3, 2) will be eighth on the path."

The manager was ecstatic. She even began an advertising campaign that used as its slogan: **Everybody is infinitely welcome.**

Professor Finkelstein-Ostrovsky-Cantorovich pointed out that the rational numbers could be arranged in the hotel in a great many different ways.

"One way is the following. We define for each fraction n/m a height that is the sum of its numerator and denominator. In other words, the height, h, of n/m is $n + m$. The lowest height would be 2, and there is only one fraction with this height, namely 1/1. Two rational numbers have the height of 3, and these are 1/2 and 2/1. The numbers 1/3, 2/2, and 3/1 all have height $h = 4$, and for $h = 5$, there are four numbers: 1/4, 2/3, 2/3, 4/1. So all the rational numbers can be arranged according to their increasing height."[1]

h	2	3	3	4	4	4	5	5	5	5	6	...
Number	1/1	1/2	2/1	1/3	2/2	3/1	1/4	2/3	3/2	4/1	1/5	...
Room	1	2	3	4	5	6	7	8	9	10	11	...

Brain-twister

Prove that in the suggested arrangement above, the number n/m will reside in the room whose number is $\frac{1}{2} \bullet (n + m - 2)(n + m - 1) + n$.

For example, number 2/3 ($n = 2$, $m = 3$) will be in room $\frac{1}{2} \bullet (2 + 3 - 2)(2 + 3 - 1) + 2 = 8$.

Hint: $1+2+\cdots+n=\frac{n(n+1)}{2}$

The hotel's fame as a place that could accommodate any group of guests became known far and wide. It didn't matter if the group that arrived was finite or infinite, it didn't matter if the hotel was already occupied or not, and it didn't even matter if the hotel was booked solid. As soon as a new group of guests arrived, places could be found for them all.

One day, something happened that Omega hadn't imagined. On the morning of that day, an email arrived from distant star Delta Continua with the message that all the numbers between 0 and 1 were interested in coming for a visit. Our manager was, of course, aware of the fact that there are "not a few" numbers between 0 and 1, such as, $\sqrt[3]{3/2}$, $e^6-\pi-\pi^5$, 1/2, 3/156, $e/47$, $(5 + 13\sqrt{2})/213$… Nevertheless, she didn't imagine any problem finding them all accommodation. Hadn't infinitely many infinite sets already stayed at the hotel? What problems could possibly arise as a result of having to accommodate one single infinite group?

Nevertheless, a problem there was, and all her attempts to solve it came up null. She had no alternative but to again turn to Professor Finkelstein-Ostrovsky-Cantorovich or Sigma or Lambda for help. Omega decided to call the professor. To her surprise and disappointment, not only could the eminent professor not provide a solution, he actually determined that this problem *had* no solution.

"What if I evict the natural numbers from the hotel? Could that help matters?" asked Omega, trying nevertheless.

"Nope," answered the professor assuredly.

"How is it possible that an infinite hotel that is empty

won't have enough room for one group of guests?" Omega did not want to acquiesce to the bad news.

"Don't be so stubborn. Instead of searching for a way to accommodate the numbers," suggested the professor, "let me prove to you that not only can an infinite hotel not find places for all the numbers between 0 and 1, but cannot even find places for all the numbers that are written only with the digits 0 and 1."

"Are you serious?" asked the manager.

"Professor Finkelstein-Ostrovsky-Cantorovich is always serious when he is talking about mathematics or music," he said, talking about himself in the third person.

"Okay. So explain yourself." And Omega prepared herself to hear the explanation.

Professor Finkelstein-Ostrovsky-Cantorovich explains

"The thing that we must determine first is that we will use infinite notation for every number. What I mean is that instead of writing 0.101, we shall write 0.101000.... Now, let us first assume that we will be successful in our mission and we will be able to find a room in the hotel for all the numbers."

"I'm guessing you are about to show me a proof by contradiction, aren't you? This is so typical of a mathematician!" said Omega.

"Here is the arrangement according to rooms: A_1 will be in room 1, A_2 will be in room 2, A_3 in room 3, and so forth. And who are these 'A' people? Well, we will know them because they will look like the following with their 'new' names:

$$A_1 = 0.a_{11}a_{12}a_{13}a_{14}a_{15}\ldots$$
$$A_2 = 0.a_{21}a_{22}a_{23}a_{24}a_{25}\ldots$$
$$A_3 = 0.a_{31}a_{32}a_{33}a_{34}a_{35}\ldots$$
$$A_4 = 0.a_{41}a_{42}a_{43}a_{44}a_{45}\ldots$$
$$A_5 = 0.a_{51}a_{52}a_{53}a_{55}a_{55}\ldots$$
....

"In other words a_{ik} will be the *k*th digit after the decimal point for the number that will be residing in room *i*. Remember, too, that every digit denoted by a_{ik} is either a 0 or a 1. I'll demonstrate. Let us assume that 0.111000110010... is in room 3. So, for this number, $a_{31} = 1$, $a_{32} = 1$, $a_{33} = 1$, $a_{34} = 0$, $a_{35} = 0$ ... and the rest is obvious."

The professor continued: "Now, I can show you a number that can be found between 0 and 1, that is to say, it is a member of this group of guests who have come from Delta Continua, but who won't be among the numbers staying at the hotel. This fact will prove that it is impossible to find a place for all the numbers between 0 and 1 in the hotel, because the list we compiled is too general.

"We shall name the number that is *not* in the hotel 'B', which will be written, of course, as $B = 0.b_1b_2b_3b_4\ldots$, and where b_i is either 0 or 1 and formed in such a way as to ensure that b_i will not be equal to a_{ii} (a_{ii} are the numbers that appear on the diagonal of the list we compiled). How do we do this?

"The idea is extremely simple. If $a_{ii} = 0$, then b_i will be 1. On the other hand, if $a_{ii} = 1$, than b_i will be 0.

"Come and I will show you an example. Let us assume that we arranged all the numbers between 0 and 1 that are written only with 0s and 1s. The arrangement is entirely arbitrary, but let us assume that arrangement is as follows:

$A_1 = 0.\mathbf{0}10010001\ldots$
$A_2 = 0.0\mathbf{1}0101010\ldots$
$A_3 = 0.11\mathbf{0}110110\ldots$
$A_4 = 0.100\mathbf{1}10111\ldots$
$A_5 = 0.0111\mathbf{1}1110\ldots$
....

"Now we will form B. The b_1 digit will be defined as 1 because $a_{11} = 0$ (the first digit after the decimal point in A_1 is 0); b_2 will be defined a 0, because $a_{22} = 1$ (a_{22} is the second digit in A_2); digit b_3 will be defined as 1 because $a_{33} = 0$. And so forth."

"But how do you know that B is not in the hotel?" Omega could not restrain herself.

"It's quite obvious. The first digit in B after the decimal point, b_1, will not be the same as the first digit after the decimal point in A_1 (that is the digit in the a_{11} place). We know this because we specifically constructed B to have the *other* digit in that place. So, obviously, B cannot be equal to A_1 even if each and every one of its other digits is exactly the same as all the other digits in A_1.

"Now, we come to the second digit after B's decimal point, that is to say, b_2. This particular number will be different than the second digit in A_2 for the very same reason, and therefore, no matter what the other digits that make up B are, B cannot be exactly the same as A_2.

"We continue along for each and every digit of B – all infinite of them. The result will be the same for each. There will always be at least one digit that is different from the numbers that make up the A_i group. Thus we must conclude that the number B cannot be equal to any particular A. That is to say, B will not be among the guests of the hotel. He arrived with all his friends from Delta Continua, but he is not staying at the hotel like they are."

"If that is the case, I will add B to the top of the list, before A_1!" Omega jumped up and down in excitement at the new idea that had just popped into her head. (Omega didn't make it difficult for the professor, but she did manage to annoy him.)

"You simply didn't understand my explanation at all! Look, even if you add B to the list, I will always be able to form some new number that won't be on the list – let's call it Y – in exactly the same way that I formed the number B."

"You're right. But I still don't understand how it is possible that any guest won't find a place in an infinite hotel."

"What this means is that even though the number of rooms in your hotel is indeed infinite, the number of guests that want to stay here is a number that is even *more* infinite," explained the professor.

"What are you talking about? *More* infinite?" asked Omega with great agitation. "Explain to me how there is an infinite that is more infinite than infinite!"

But the old professor was exhausted from his long explanation.

Intermezzo

A finite–infinite dilemma: From googol to Google

How does one compare the size of different sets? Is the number of drops of water in the Atlantic Ocean greater than the different possible arrangements on a chessboard? Is the number of tunes that one can compose greater than the number of rational numbers between 0 and 1? How can we know that in one set there are fewer elements than another when we are talking about sets that are extremely large or infinite? When can we say that two sets are equal in size? Is it easy to distinguish between a very large group and an infinite one? Even the great Archimedes was already aware of the problems that can crop up as a result of confusing an extremely large set with an infinite one. *The Sand Reckoner* is a work by the great man from Syracuse in which he decided to find an upper bound for the number of grains of sand in the universe.

> *There are some who think that the number of the sand is infinite in multitude; and I mean by the sand that which is found in every region whether inhabited or uninhabited. Again there are others who, without regarding it as infinite, yet think that no number can ever be named which is greater than the number of grains of sand.*
>
> (Archimedes)

Archimedes found the limit to the number of grains of sand in the universe to be 10^{63} and through this he proved that the two previous claims were false.

Many years after Archimedes finished his calculations, in 1938 to be precise, the English astrophysicist, astronomer and mathematician Sir Arthur Stanley Eddington delivered a lecture at Trinity College in Cambridge and stated:

> I believe there are 15,747,724,136,275,002,577,605, 653,961,181,555,468,044,717,914,527,116,709,366, 231,425,076,185,631,031,296 protons in the universe and the same number of electrons.

This huge number is known today as the "Eddington number". Though it looks quite impressive, it has nothing to do with infinity.

In their book *Mathematics and Imagination* (1940), mathematicians Edward Kasner and James Newman wrote that one must understand that "a great many" and "infinity" are two totally different concepts. There is no point at which a large star changes to infinite. We can write a number that is as large as we wish, and it won't be any closer to infinity than 1 or 7 are.

An interesting morsel of trivia is that the term "googol" first appeared in the above-mentioned book. Kasner's nephew, Milton, who was then nine, suggested that name for a number that is written with a one followed by a hundred zeroes. Interestingly, the name "Google" came about as a misspelling of googol.

That same strange boy also suggested the term "googolplex" for the number that is made up of a one and a lot of zeroes: "Googolplex should be 1, followed by writing zeroes until you get tired." Today, googolplex represents a number that is defined much more accurately as 10^{googol}. Don't even think to try to imagine this number. Astronomer and writer Carl Sagan (1934–1996) pointed out on his television show *Cosmos: A Personal Voyage* that it is impossible to write the number googolplex due to a very serious problem: there is not enough room in the observed universe for all its digits.

Yet, even so, googolplex is an infinite distance away from infinity. In fact, it too is no closer to infinity than 1 or 7 or any other number you might want to name.

Even the number that is googolplex to the power of googolplex is still definitely finite. I'm going to call this number Poohplex in honour of my dearest friend, the cute, plump Winnie the Pooh. So if googolplex is beyond all human imagination, what can you say about Poohplex? You can invent numbers that are as enormous as you like, and even give them any names you like. You could have Poohplex to the power of Poohplex, and then think of the factorial of that result – I get a headache just trying to comprehend the size of such numbers – and no matter what, these are all finite numbers and no closer to infinity than 7.

Let's get back to matters of infinity.

LESSON 8

CARDINALS AND THE TAMING OF THE INFINITE

About football players and fashion models (1:1 correspondence)

We return to the question: when can two sets be deemed equal in size?

In the case of finite sets, this is no problem. At least, in principle, one can solve the problem by counting: if set A has the same number of elements as set B, we can say that both sets are equal in size.

However, a problem arises when we start talking about infinite sets. In this case, it is impossible to reach a final count. Do you think you can compare the size of two groups without counting? As it turns out, you can.

Let us begin by understanding a bit about a different idea of comparing finite sets. Imagine a fashionable club where the annual get-together of top fashion models and famous football players is well underway. The party is in full swing and many football players and lots of models are boogying on the dance floor.

Can we discover, without counting, whether there are more football players than models, more models than football players, or if perhaps the number of each is equal?

The solution to this problem is pretty straightforward: all one has to do is get some soft music going and announce that every football player must ask a model to dance. At this point, there are three possibilities:

1 Everyone will be dancing, that is to say, the number of football players is equal to the number of models.
2 There are some football players who can't find themselves a partner and are standing, sad and lonely, at the bar. In this case, it is clear that there are more football players than models.
3 There are some models who aren't dancing – the set of models is greater than the set of football players.

It is important to point out that this method of comparison does not leave us knowing the precise number of football players and models. However, we wanted to *compare* the size of each group, and this is exactly what we did.

This comparison method also works for comparing infinite sets, for which counting is impossible.

At this point, it's time to become familiar with two boring – yet essential – concepts.

One-to-one, or injective, correspondence or mapping (1:1)

Correspondence between members of set A and set B such that different elements of set A will correspond (pair) to different elements of set B, and vice versa, is called one-to-one mapping (*injective*). The short-form notation for this type of correspondence is 1:1.

For example, let us assume that there are three football players – Ronaldo, Messi, and Mbappé – and four models – Adrianna, Giselle, Kate, and Nina. If we match the football players with the models like this:

A	B
Ronaldo	Nina
Messi	Giselle
Mbappé	Kate

we have 1:1 correspondence between the two sets we made because any two football players (the different elements of set A) are paired with different models (the different elements of set B). The fact that Adrianna is not matched with a partner does not make a difference from the point of view of the definition. As long as each member of A has a unique partner, there is 1:1 correspondence.

On the other hand, we might pair them up as follows:

A	B
Ronaldo	Nina
Messi	Kate
Mbappé	Kate

In this case, the correspondence is not 1:1 because there are two different football players who have been paired with the same model, Kate. There are more elements in A than there are in B.

Surjective correspondence

When there is correspondence of the elements of set B to those of set A such that for each element in B there is *at least* one corresponding element in set A, the correspondence is termed *surjective*. Note that more than one element of A may be mapped onto one element of B (in which case, this results in a mapping which is not 1:1). In this case, one says that A is surjective (*onto*) to B.

For example, let us say we now have five football players: Ronaldo, Messi, Mbappé, Kane, and Neymar, and the same four models – Adriana, Giselle, Kate, and Nina. We can give them surjective correspondence as follows:

A	B
Ronaldo	Nina
Messi	Giselle
Mbappé	Kate
Kane	Kate
Neymar	Adrianna

This is surjective because every element in B (four models) has been paired with *at least* one element in set A (football players). Note here that two football players have been "mapped onto" one of the models (Kate).

On the other hand, the following is *not* surjective correspondence:

A	B
Ronaldo	Kate
Messi	Kate
Mbappé	Adrianna
Kane	Nina
Neymar	Adrianna
	Giselle

Why? Because one of the elements in B (Giselle) has not been paired to any element in A. (Note here that two of the models have two football players each "mapped onto them", leaving poor Giselle all on her own.)

If two sets, A and B, have both 1:1 and surjective correspondences between them, it means that the elements of the sets can be arranged in "perfect" pairs – each different element in set A can be matched with an element in set B, and also that every element in set B can be matched with an element in set A. Correspondence that is both injective and surjective is called *bijective*.

It is totally convincing that, if sets A and B are both finite sets, having both 1:1 and surjective correspondences is possible only provided that both sets have the same number of elements. I shall explain – 1:1 correspondence (bijective) means that there are an equal or greater number of elements in set B than in set A, whereas surjective correspondence implies there is an equal or greater number of elements in set A (every element in B can be associated with several elements of A). Both properties together assure that if A and B are finite sets, then the number of elements in each must be the same.

One can successfully demonstrate that the correspondence between a set of football players and a set of models is both 1:1 and surjective if, and only if, both sets have the same number of elements, as below (for nostalgic people).

A	B
Zidane	Claudia
Pelé	Cindy
Maradona	Kate
Beckham	Naomi
Maldini	Linda

Here is another example:

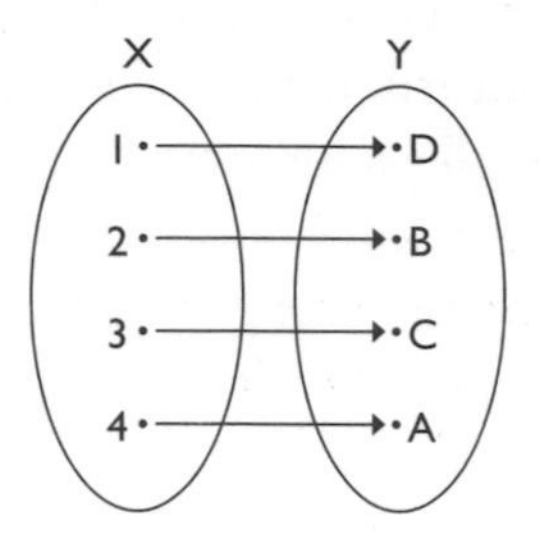

This also demonstrates 1:1 and surjective correspondences, and we didn't even have to involve any football players or models.

Now that we have that cleared up, let us get back to infinite sets. Based on the foregoing, it seems natural to define equality between the number of elements in two sets (finite or infinite) as follows:

DEFINITION: EQUALITY OF CARDINALITY

Two sets, A and B, are of equal cardinality if there is some (any) correspondence between the elements of set A and the elements of set B is both 1:1 (injective) and surjective.

What is this "cardinality"? You may have remembered we touched upon it a while back. For finite sets, the meaning of cardinality is clear.

DEFINITION: CARDINALITY OF FINITE SETS

Regarding finite sets the notion of cardinality is just a fancy term for "number of elements of the set". For example, A = {17, 42, 1729, 1,234,321} has four elements, and therefore "4" is its cardinality. Henceforth, we shall write it as follows: #A = 4.

For infinite sets, however, the notion of "number of elements in the set" is not and *cannot* be obvious. All we can do with infinite sets is *compare* their cardinalities.

Galileo Galilei's paradox

At the beginning of the seventeenth century, Galileo Galilei presented the paradox that bears his name. Galileo's paradox involves the 1:1 and surjective correspondences between the set of natural numbers {1, 2, 3, 4, …} and the set of squares {1, 2, 4, 9, 16, …}. It is possible to pair the sets as illustrated below. It should be clear that every element in A has one and only one element that matches it in B, and vice versa:

The paradox involved here is that, on one hand, we have equality of cardinality (that is to say, both 1:1 and surjective correspondence) between the set of natural numbers and a proper subset, namely, a subset which is not equal to the set itself – in our case, the set of squares. How is this possible if there are more natural numbers than squares, meaning that there are more elements in one set than the other? How is it possible that they are equinumerous!!?

Set A	Number	1	2	3	4	5	6	7	8	9	10	...
Set B	Square	1	4	9	16	25	36	49	64	81	100	...

Georg Cantor

Galileo Galilei

DEFINITION: PARADOX

A tenet or proposition contrary to received opinion; an assertion or sentiment seemingly contradictory, or opposed to common sense; that which in appearance or terms is absurd, but yet may be in fact true.

How wonderful that we have met with a paradox!
Now we have some hope of making progress.

(Niels Bohr)

How I agree with Niels Bohr! Paradoxes are such wonderful things to shake up the thinking process.

Galileo believed that this paradox, which he wrote about in his book *Dialogues on Two New Sciences*, proved that when talking about infinite sets, one cannot use adjectives like "equal", "smaller", or "larger", and that, in fact, as we quoted much earlier in this book, it is preferable for those with finite brains to steer clear of anything to do with the infinite.

But so what if we have finite minds? Why should we be shackled and bound to finite thinking schemes? It was Cantor and Dedekind who sought to turn this apparent problem into the beginning of a new theory.

DEFINITION: SUBSET

Set A may be called a subset of set B if *every* element of set A is a member of B.

For example:

A = {Gustav Mahler, Gustav Holst, Gustavo Dudamel}

B = {Gustav Mahler, Gustav Klimt, Gustav Holst, Gustavo Dudamel, Gustave Doré, Gustavo Boccoli, Gustave Courbet, Hurricane Gustav, Gustaf V of Sweden}

Set A is a subset of set B because every element in set A is also in B. The definition also leads to the observation that any set will be a subset of itself.

Let us re-examine Galileo's paradox. But before we do, there are a couple of definitions that we must learn.
As a reminder:

DEFINITION: PROPER SUBSET

If set A is a subset to set B but is *not equal* to set B, we say that A is a proper subset of B. In the example above, A is a proper subset of B.

THE CANTOR–DEDEKIND DEFINITION OF INFINITE SETS

A set is called infinite if there exists both 1:1 (injective) and surjective correspondences between it and at least one of its proper subsets. (Just to remind you, with finite sets, a proper subset of set A cannot have a 1:1 correspondence with A!)

For example, the set of natural numbers is infinite because, as Galileo demonstrated, it is equivalent to a proper subset of itself – the set of squares. If we wish to use the fancy word we just learned, we can say that the set of natural numbers and the set of squares have the same cardinality. What is important to remember is this: with finite sets, the statement "the part is always smaller than the whole" holds; with infinite sets, this statement does not hold. We have observed more than enough validation for this: Galileo Galilei's paradox, Russell's version of "Achilles and the tortoise" (see "Zeno's apology" above) from Zeno's school, all the wonders of Hilbert's infinite hotel ...

Brain-twister: ***Paradiso e inferno***

A man is condemned to eternal suffering in hell. Another is to spend eternity in paradise. One day a year, they switch: the miserable one is allowed to enjoy the delightful freshness of paradise, and the joyful resident of paradise gets a taste of the horrors of hell.

Mathematically speaking (calculating cardinalities), is there a difference between these two ways of spending the ever-after?

If you say there is a difference, explain why.

If you say there is no difference, answer the following: where would you choose to spend eternity?

Galileo's paradox is not a paradox any more; it simply transformed itself into proof of the infinite nature of the natural numbers. Obviously, one can find many other subsets that will be equinumerous – that is, will have the same cardinality – as the set of natural numbers (the set of prime numbers, the set of even numbers, the set of natural numbers that are divisible by 101, the set of numbers that are exact factorials {1, 2, 6, 24, 120, 720, 5,040, 40,320, 362,880, 3,628,800...} and so on).

The cardinality of infinite sets

Set D = {1, 2, 3, 4, 5}. It is not, by definition, infinite. Why? Because if we take a proper subset from it, E, we will not be able to find both 1:1 and surjective correspondences between the two sets. In other words, we would be unable to arrange all the different elements of D and all the different members of E in pairs.

As mentioned above, the cardinality of a finite set is simply the number of elements it has, therefore, we can write #A = n.

But how do we denote the cardinality of infinite sets? We can't *count* the elements in infinite sets!

Do infinite sets even have a cardinality?

If they do, are there some infinite cardinalities that are greater than others (a visit to the infinite hotel gives more than a hint that this is possible)?

Is there a "lowest" infinite cardinality?

Is there a "greatest" infinite cardinality? Is infinite cardinality "infinite"? If so, how would we denote such a cardinality value?

For the answers to these and other questions, stay tuned!

Countably infinite sets arrive at Hilbert's hotel

Every finite set is, obviously, a countable set. You begin with the first element, move to the next, etc, and, at some point (even if it has a googolplex of members), you (or your descendants) will reach the last element. In fact, an infinite set is defined as a "countably infinite" set if it has the same cardinality as the set of natural numbers, meaning that it has both 1:1 and surjective correspondence with the set of natural numbers. In other words, its members can be arranged sequentially and thus its members can be accommodated in Hilbert's hotel in some fashion. They are countable in the sense that we can arrange them in such a way that we have the first element, then the second element, then the third ..., and although this process will never terminate, we are still counting the elements! Therefore we use the word "countable".

We have already seen that there was no problem accommodating the set of integers and the set of rational numbers as guests in Hilbert's infinite hotel. This means that these two sets are undoubtedly countable.

I will remind you of the arrangement of the rational numbers in the hotel. Recall that we arranged them by increasing "height", where the "height" of fraction a/b was defined as $h = a + b$, and where fractions with identical heights were arranged according to the increasing value of the numerator (see the chart below). It is clear that the fractions, as arranged below, show an injective correspondence with the natural numbers.

h	2	3	3	4	4	4	5	5	5	5	6	...
Rational number	1/1	1/2	2/1	1/3	2/2	3/1	1/4	2/3	3/2	4/1	1/5	...
Room #	1	2	3	4	5	6	7	8	9	10	11	...

David Hilbert

Emmy Noether

Here's a reminder of how the integers could be arranged in the most luxurious mathematical hotel in the universe:

Guest (Z)	0	1	−1	2	−2	3	−3	4	−4	...
Room #	1	2	3	4	5	6	7	8	9	...

We have already noted that the cardinality of a finite set is denoted by #A. However, because one can never finish counting an infinite set, no value for "*n*" can exist for its cardinality. Therefore, the cardinality of a countably infinite set must be denoted otherwise. Cantor used the symbol $\aleph_0$ (aleph-naught or aleph-zero or aleph-null). It is the Hebrew letter "aleph" with a subscript of zero.[1] Assuming that N represents the set of natural numbers and Z represents the set of integers (all positive and negative numbers, and zero), then one can write both $\#N = \aleph_0$ and $\#Z = \aleph_0$.

Representing the cardinality of a countably infinite set by $\aleph_0$ hints that $\aleph_0$ is probably the lowest infinite cardinality, and that there may be greater cardinalities for infinite sets (which we can't count anyway!). This, in fact, is true.

Mini brain-twister

Prove that every infinite set includes a countable infinite set.

What this exercise implies, among other things, is that an infinite set *may* fill up the infinite hotel. The emphasis is on "may".

For example, the set of numbers divisible by 3 can be housed in the infinite hotel in such a way that they don't fill up all the rooms simply by having each stay in a room with its own number.

Room	1	2	3	4	5	6	7	8	9	10	11	12	...	...
Guest			3			6			9			12		

There are infinite rooms that remain vacant.

But if every divisible-by-3 number is assigned to a room whose number is one-third of its value, the hotel will be filled completely.

Room	1	2	3	4	5	6	7	8	9	10	11	12	...	...
Guest	3	6	9	12	15	18	21	24	27	30	33	36	...	...

This shows us that the set of numbers divisible by 3 is a countably infinite set. (Because there is a bijection between it and the natural numbers, as you can see in the table.)

The set of numbers divisible by googolplex is also infinite and also countable, as is the set of numbers divisible by Poohplex. Try to imagine how many numbers one must pass until we meet up with Poohplex! Then we will have to pass the same quantity of numbers until we get to 2 × Poohplex! Nevertheless, the cardinality of the numbers divisible by Poohplex is equal to the cardinality of numbers divisible by 21, or that of even numbers, and that of natural numbers.

The cardinality of all these sets is $\aleph_0$.

Believe it or not!

Our pale reasoning hides the infinite from us.

(Jim Morrison, The Doors)

Algebraic numbers vacation at Hilbert's hotel

Our expedition to Hilbert's hotel demonstrated that not every set can stay there, despite the fact that this is an *infinite* hotel. The quantity of the set of *all* the numbers between 0 and 1 was too great for all of them to stay at the hotel.

These numbers are *not* countably infinite, since there is not 1:1 and surjective correspondence between them and the natural numbers. Are there any other sets of numbers that are infinite but not countable, that is to say, the type that the infinite hotel cannot accommodate?

An interesting set of this type is the set of non-algebraic numbers, which will be defined in a moment. But first, we must be clear on what an algebraic number is.

Recall that a rational number is a number, q, that can be written as the ratio of two integers $\frac{c}{d}$.

Equivalently, we can define q as a rational number if and only if it is the solution for a "first degree" equation, namely, an equation in the form

$$ax + b = 0$$

where the coefficients a and b are integers.

It is clear that every rational number $q = \frac{c}{d}$ satisfies $dq - c = 0$ and therefore is a solution to the first degree equation $dx + (-c) = 0$. For example $\frac{19}{77}$ is the solution for the equation

$$77x - 19 = 0$$

What exactly, then, is an algebraic number?

DEFINITION: AN ALGEBRAIC NUMBER

A number is considered algebraic if it is a root (i.e. solution) of an equation of the form:

$$a_n x^n + a_{n-1} x^{n-1} + \ldots + a_1 x + a_0 = 0$$

where every a_k is an integer.

A number that is not an algebraic number is termed "transcendental".

The left-hand side of the equation above is called a polynomial of degree n, provided that a_n is not 0.

From the definition, it is immediately clear that *all rational numbers* are algebraic. Yet, there are algebraic numbers that are irrational.[2] Here are some examples:

$\sqrt{2}$ is an algebraic number because it solves the equation $x^2 - 2 = 0$.

The third root of 3/2 ($\sqrt[3]{3/2}$) is an algebraic number because it solves the equation $2x^3 - 3 = 0$.

$\sqrt{-1} = i$ is an algebraic number (but not a real number) because the root of $x^2 + 1 = 0$.

The golden ratio, φ, is an algebraic number because it solves the equation $x^2 - x - 1 = 0$.

In short, there are "numerous" algebraic numbers, because there are "numerous" equations of the polynomials $a_n x^n + a_{n-1} x^{n-1} + \ldots + a_1 x + a_0$.

Based on the above, this next proposition might look a bit surprising:

Theorem: The set of algebraic numbers is countable.

Proof: Let us examine the equation

$$a_n x^n + a_{n-1} x^{n-1} + \dots + a_1 x + a_0 = 0$$

We assume that a_n is positive. If this were not the case, we would multiply the equation by (–1) and the resulting equation will have the same roots.

Similar to the arrangement of rational numbers in the hotel, we define a height, H, for every polynomial.

$$H = n + a_n + |a_{n-1}| + |a_{n-2}| + \dots + |a_1| + |a_0|$$

(The symbol $|m|$ denotes the absolute value of a number. If the number is positive, its absolute value is equal to itself: $|37| = 37$. If the number is negative, its absolute value switches to positive $|-234| = 234$.)

Now, we can write all the equations (some will have no solution) in order of their height.

For example, for H = 1 there is only one polynomial, and that is simply the polynomial 1, which is independent of x, and would yield the equation $1 = 0$, which has no solution. This is not a valid equation, and does not give us any algebraic number.

For H = 2, we have two equations: $x = 0$ and $2 = 0$. The first one will give us the algebraic number 0 and the second one, again, is an invalid equation and has no roots.

For H = 3, we get the following equations: $3 = 0$, $x - 1 = 0$, $2x = 0$, $x + 1 = 0$, and finally, $x^2 = 0$. From all these equations, the first yield no algebraic numbers, and from the others we get only two new algebraic numbers: 1 and –1.

Okay, I hope that the main idea is clear.

When we get to H = 5, we will deal with √2 (check it). At each height there are a finite many equations and each equation has finitely many solutions, consequently at each height we add finitely many algebraic numbers. This proves that the set of algebraic numbers is in fact a collection of countably many finite sets. Hence, there is no problem accommodating all the algebraic numbers in Hilbert's infinity hotel. This also means that the set of algebraic numbers is a countably infinite set and its cardinality is $\aleph_0$.

It's rather unbelievable, but the cardinality of the numbers that are divisible by Poohplex in the power of Poohplex is equal to the cardinality of the set of algebraic numbers.

ℵ: A greater infinity – The cardinality of the continuum

It is no problem to prove a set is countable. One just needs to find a 1:1 and onto (surjective) correspondence with the set of natural numbers. The root of the problem is this: in order to prove that a particular set is countable, it is sufficient to show that its elements can be placed into some sequential order, but in order to *prove* that a set is non-countable, we must prove that there is *absolutely no way* to arrange its elements sequentially. (This is similar to the "problem" of proving there is at least one ant in a room, compared to the problem of proving that there is definitely no ant anywhere in a room. Once you have found an ant, you've proved the first, but not finding an ant up to a specific moment does not mean you won't find one later.)

As I have already pointed out, Georg Cantor, in 1891, came up with a sophisticated method called "Cantor's diagonal method" that can aid in proving the impossibility of counting a quantity of varied objects. We already encountered this method when Professor Finkelstein-Ostrovsky-Cantorovich proved that it would be impossible to accommodate the numbers between 0 and 1 whose decimals numbers consist only of 0 and 1 in the infinite hotel. This exact method can be used to prove that the set of *all* the numbers between 0 and 1 are uncountable (prove it!). This is pretty much expected, since the "numbers-between-0-and-1-written-in-decimal-using-only-0-and-1-digits" set is a proper subset of the set of all the numbers between 0 and 1. In addition, if we remember that every existing number can be represented in binary notation using only the 0 and 1 digits, we might become easily convinced that both sets are equivalent with respect to their cardinality. (Why?)

Sets of numbers that are infinite and that can't be ordered sequentially are called – not surprisingly – uncountable. The set of all the points on the number line between 0 and 1 is uncountable, so its cardinality is not $\aleph_0$. Therefore, in order to denote the cardinality of all the real numbers (or any segment on the line of real numbers), we need a new symbol! For this we use $\aleph$. We say that $\aleph$ is the cardinality of the continuum. Note, however, that uncountable sets are not necessarily of cardinality $\aleph$.

Paroles, paroles, paroles

Because the concept of Cantor's diagonal is not only beautiful but also important, I will explain it again, this time by showing why the set of all the words of infinite length that are made up using only letters a and b cannot be counted. That is to say, the cardinality of such a set is uncountable.

If you have already understood Professor Finkelstein-Ostrovsky-Cantorovich's explanation to Omega about the decimal numbers, you should have no problem with this. It is exactly the same, just a different example. If you didn't entirely get his explanation, hopefully you will this time.

The proof will be done as a proof by contradiction, that is to say, we will assume the opposite hypothesis: all the words can be arranged sequentially. Then we shall discover that this leads to a contradiction, meaning that our opposite hypothesis is false.

Here is the arrangement of the words:

A1	=	**a**	b	a	b	b	b	b	b	b	b	...	...	...
A2	=	a	**b**	b	a	a	b	b	a	a	b	...	...	...
A3	=	b	b	**a**	b	a	b	a	a	b	a	...	...	...
A4	=	a	a	a	**a**	b	b	b	b	a	a	...	...	...
A5	=	b	a	a	a	**a**	a	a	a	b	a	...	...	...
A6	=	a	a	a	a	b	**a**	a	a	a	a	...	...	...
A7	=	b	b	b	b	b	b	**b**	b	b	b	...	...	...
A8	=	a	a	b	b	a	a	a	**a**	b	a	...	...	...
A9	=	a	a	a	a	b	b	b	b	**a**	b	...	...	...
...													

Based on Cantor's diagonal method, and similar to how we worked with the numbers between 0 and 1, we shall now form a word, A0, which is not already included anywhere in the set depicted above no matter how the arrangement is made. Take a good look at the diagram above, and note the highlighted letters that appear on the diagonal. The new word, A0, will be constructed as follows: The first letter will be different from the first letter in A1 (the first letter in A1 is a, so we will use b); the second letter will be different from the second letter in A2 (which is b, so we use a); the third letter will be different from the third letter in A3 (this time we will use b); and so on and so forth.

So, here is our new word: A0 = babbbbabb…

I will leave it my wise readers to convince themselves that there is no way that A0 can possibly appear (that is to say, be a repeat of any word) in our original infinite list since it is necessarily different from the A*i* words at least with respect to the letter that appears in the *i* place.

And again, as Omega discovered, adding A0 to the list does not change a thing, because we will always be able to repeat the process and build another word, let's call this one Aℵ, which will be different from each and every word in the infinite list we compiled. To conclude, the set of all the words of infinite length that are composed only of the letters a and b have the cardinality of the continuum.

Clearly, the set of all the words of infinite length that are composed using three different letters (not just a and b), or four, or five (or any other number) different letters will also have uncountable cardinality, which does not automatically mean that it has the cardinality of the continuum. However, since we can construct 1:1 and surjective mapping of this set onto the set of numbers with 0 and 1, we see that its cardinality is, indeed, ℵ.

Another proof (a nice one) that all the numbers on the segment [0,1] are uncountable

Let us assume the opposite hypothesis: all the points on line [0,1] can be counted. What this implies is that they can somehow be arranged in a sequential order – {p_1, p_2, p_3, p_4 …}. To prove (or disprove) this hypothesis, we begin by surrounding the centre point p_1 with an interval that is, say, of size 1/10, p_2 within an interval that is 1/100 long, p_3 within an interval that is 1/1,000 long, and so on and so forth. Now, as all the points in the interval [0,1] are in at least one of these intervals (recall that {p_1, p_2, p_3, p_4 …} was an enumeration of *all* the numbers between 0 and 1), we end up with a set that covers the line [0,1]. On the other hand, we could add together the lengths of all the intervals. According to the formula for infinite convergent geometric series:

$$a_1 + a_1 q + a_1 q^2 + \ldots = a_1 / (1 - q)$$

$$1/10 + 1/100 + 1/1{,}000 + 1/10{,}000 + \ldots$$
$$= (1/1)/(1 - 1/10) = 1/9.$$

We have succeeded, as it were, in covering all the points on line segment [0, 1] with intervals whose cumulative length is equal to only 1/9. However, this is clearly impossible since the length of the original line segment is 1.

We have thus reached a conflict.

Conclusion: it is impossible to write all the points between 0 and 1 as a sequence. In other words, this set is uncountable.

Since the rational numbers form a countable set, it is possible to arrange all the rational numbers that can be found on segment [0,1] in order. How? By surrounding them with intervals in such a way that the cumulative length of the segments is not more than 1/9. What this means is that the rational numbers make up, at most, 1/9th of all the numbers that exist between 0 and 1.

However, we can improve on this upper bound.

Let us assume now that the arrangement of rational numbers found on line segment [0,1] is $\{q_1, q_2, q_3, q_4, \ldots\}$. Now, we surround q_1 with an interval of length 1/1,000, q_2 with an interval of length 1/10,000, q_3 with one 1/100,000 and so on and so forth. Now, the cumulative length of all the intervals is equal to $\frac{1/1000}{1-1/10} = \frac{1}{900}$.

Obviously, we can continue reducing the length of the cumulative interval that covers all the rational numbers on [0,1], and make the cumulative length as short as we wish. A set that can be covered by a countable union of intervals whose cumulative length is shorter than any value defined beforehand is called a null set.

All truths are easy to understand once they are discovered; the point is to discover them.

(Galileo Galilei)

Mathematics is the most beautiful and most powerful creation of the human spirit.

(Stefan Banach)

This is hilarious! None is more equal than the others

As I have already mentioned, the cardinality of all the real numbers – rational and irrational – between 0 and 1 is denoted by ℵ, and is called the cardinality of the continuum. There is nothing special about the segment between 0 and 1. Its length is one unit, but the cardinality of any segment is ℵ. It is easy to see that any two segments are equivalent, meaning that there is both 1:1 and surjective correspondence between any segment AB and the set of points on another segment, CD. Here is a hint as to how we can illustrate such correspondence:

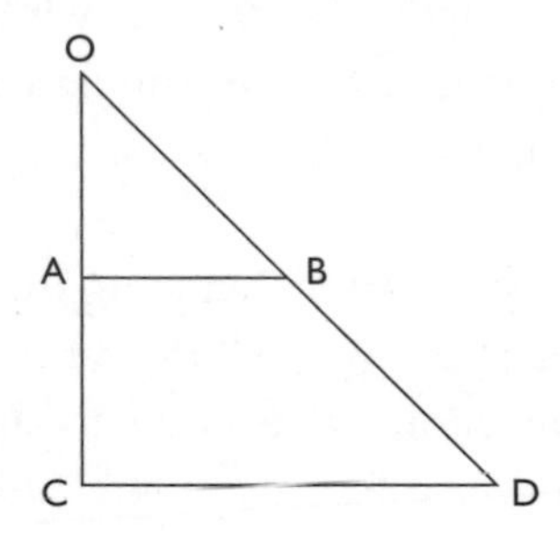

Did the hint not help? Here is the solution. For any point on AB, we can find a corresponding point on CD as shown in the illustration below.

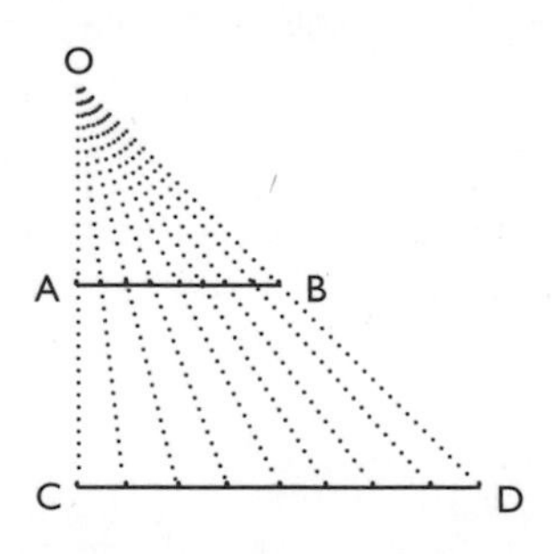

It is clear that each different point on the shorter segment, AB, can be connected to different points on segment CD. Here we have 1:1 correspondence.

Similarly, it is clear that for every point on CD, we can find a corresponding point on AB. (All one needs to do is draw a line from the point on CD to the vertex of the triangle to locate the point of intersection with AB.) This is surjective correspondence.

Since we have successfully paired all the points on both intervals of different lengths, both must have the same cardinality – ergo, the cardinality of the continuum, $\aleph$.

Here is something that is even more bizarre. It is possible (in a similar manner) to prove that the cardinality of any line segment is equal to that of an infinite ray. The figure below is a pretty broad hint. If you take a good look, I am pretty sure you will figure out how to build the correspondence between a finite segment and an infinite ray.

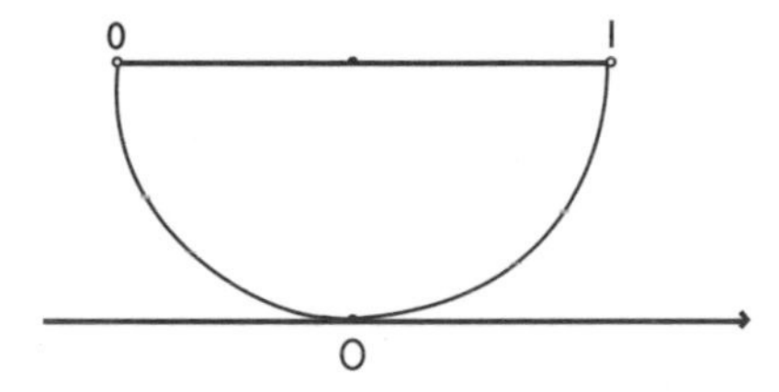

What this means is that a line segment one millimetre in length, a line segment a billion kilometres in length, and even an infinite line segment, have an equivalent "number" of points. This result might seem less astonishing if we recall that, in fact, a point has no length, no area, and no volume. Zeno would have asked how these "lengthless" points can end up forming a line that is 107 "whatevers" in length or even an infinite ray.

Moving on from mere lines, rays and segments, Cantor also proved that there is 1:1 and surjective correspondence between the points of a line segment and the points in a square or a cube!

But even more astounding and more impressive: Cantor proved 1:1 and surjective correspondence between an infinite line and between n-dimensional space (for all n!).

I'll let you in on something: this discovery was even too extreme for Cantor. This is his reaction to his discovery: "*Je le vois, mais je ne le crois pas!*" (I see it, but I don't believe it!)

And now for something completely different

You may recall that after the definition given above for algebraic numbers, I pointed out that numbers that are not algebraic are called transcendental numbers. Based on our discovery that the set of real numbers has higher cardinality than the set of algebraic ones, it seems a foregone conclusion that transcendental numbers exist, that is to say there are numbers that are *not* a root of an expression of the type $a_n x^n + a_{n-1} x^{n-1} + \ldots + a_1 x + a_0 = 0$.

But where are they? Although the concept of transcendental numbers has been around a long time, up to the nineteenth century, no one had actually "seen" one for sure.

The proof that such numbers exist did not originate with Georg Cantor. It was proven in 1844 by the eminent French mathematician Joseph Liouville. However, Cantor improved on Liouville's work by showing that transcendental numbers are the majority of the numbers. (In other words, not only are the majority of numbers not rational, the majority of numbers are not even algebraic.)

Theorem: The set of transcendental numbers is uncountable.

Proof: The set of all real numbers can be divided into two disjoint sets: the set of algebraic numbers and the set of transcendental numbers. ("Disjoint" means that a particular element cannot be a member of both sets.)

We denote algebraic numbers by "A", transcendental numbers by "T" and real numbers by "R".

DEFINITION:

The union of two sets A and B, denoted by A∪B, is the set of elements that are in A, in B, or in both A and B.

The union of the sets A and T is all the real numbers, R. Hence, we write R = A∪T.

Get ready for the point of this story: Since the cardinality of R, the set of all the real numbers, is ℵ, it is not implausible to conclude that T must be uncountable (or less).

The reason that the set of transcendental numbers, T, must be uncountable stems from the fact that the union of two countable sets must yield another countable set.

If both A and T were countable – that is to say both the algebraic numbers and transcendental numbers were countable, then we could order A = (a_1, a_2, a_3, …) and

$T = (t_1, t_2, t_3, \ldots)$. Therefore, the union $T \cup A$ would also be countable, since it could be ordered as follows:

$$T \cup A = (a_1, t_1, a_2, t_2, a_3, t_3, \ldots)$$

However, R, as you know, is uncountable. Since we know that A is countable (see the above section entitled "Algebraic numbers vacation at Hilbert's hotel") and $T \cup A = R$, it is impossible that T is also countable.

Well, if the quantity of transcendental numbers is so great that they are uncountable, it seems that we should have no problem finding an example of a transcendental number. Indeed, mathematicians should be tripping over them left and right.

But do they? In fact, they don't. Barely any transcendental numbers have even been identified!

Let us try. Perhaps $(\sqrt{2} + \varphi)$ is transcendental? Nope. It turns out that this number is algebraic – you can try forming the polynomial equation (with whole coefficients) for which this number is a solution for. In fact, I dare you to try to find any number that is not algebraic.

So what do we have here? We have proven that the number of transcendental numbers is not only infinite, but even uncountable. The problem is that this is a proof of existence and not a constructive proof. In other words, while the proof might convince us that there are infinite transcendental numbers (even due to the cardinality of the continuum), it does not give even the tiniest hint of *how* to find even one such number.

In 1844, as we mentioned, Liouville discovered a transcendental number. Here it is:

$$L = \sum_{f=1}^{\infty} 10^{-j!} = 0.110001000000000000000001000\ldots$$

It may be unclear to you exactly what this number is, so let me explain.

Liouville's number is constructed as follows:

Step 1: We calculate all the factorials: 1! = 1, 2! = 2, 3! = 6, 4! = 24, 5! = 120 …

Step 2: We form a number so that only zeroes and ones appear after the decimal point with the 1 digit in the 1st, 2nd, 6th, 24th, 120th … places, and the 0 digit everywhere else.

Liouville proved that this number is not the root of any polynomial with whole coefficients.

As you might imagine, the proof of this is not very simple and you will just have to trust me (and Liouville) that this is true.

Look at this next number, 3.140001000000000000000005…, which is formed in a similar way. This is the number which is obtained from the decimal notation for π where all the digits except for the 1st, 2nd, 6th, 24th, 120th … places (the places corresponding to 1!, 2!, 3!, …) digits have been replace with a zero. In the places noted, the original digits are written one after another.

(π =3.141592653589793… therefore we have the following expansion, 5 is in the 24th place, 9 will appear in the 120th place, 2 on the 720th place and so on.)

Mathematicians can prove that this is also a transcendental number (and again, you will just have to trust that they know what they are doing).

But what about π itself? Is it transcendental?

π: I am not rational!
You cannot guess what I am
going to do ...

The fact that π is irrational (that is to say, there is no fraction a/b that will give us the value of π) was already acknowledged – but not proven – by ninth-century Persian mathematician, astronomer, and geographer Muhammad ibn Musa al-Khwarizmi (his Latinized name was Algoritmi, from which originates the word "algorithm"). He was a major figure in the famous House of Wisdom in Baghdad during the golden age of the Islamic culture. He made many tremendously important contributions to the field of algebra, and, in fact, the very term "algebra" is from "al-jabr", which appears in the title of the seminal book al-Khwarizmi wrote on that subject in 820.

Maimonides also believed in – but didn't prove – the irrationality of π. Formal proof was given only in 1768 by Swiss mathematician (Euler wasn't the only one!) Johann Heinrich Lambert.

While the proof that π is an irrational number is relatively simple; proving that π is transcendental is complicated to the extreme. It took more than 100 years until German mathematician Ferdinand von Lindemann proved, in 1882, that π was transcendental, that is to say, it is not the root of any polynomial with whole coefficients.

A few years earlier, in 1873, French mathematician (have you noticed how many French mathematicians we have encountered?) Charles Hermite proved that Euler's number, *e*, is a transcendental number.[3] Proving that a number is transcendental (especially the one for π) is an especially long and complicated process, and we will not go into this here. Generally speaking, just try to imagine how one might go about trying to prove that there is a number that does not yield zero for any equation in the form $a_n x^n + a_{n-1} x^{n-1} + \ldots + a_1 x + a_0 = 0$. This is no easy task!

To demonstrate how difficult it is, I will point out that, at present, it is still not known if π to the power of π (π^π) is algebraic or transcendental. David Hilbert (I remind you that he is the proud "owner" of the infinite hotel) questioned whether $2^{\sqrt{2}}$ was algebraic or transcendental. Today we know that this number is transcendental. In fact, proof of this is part of a general theorem called the Gelfond-Schneider theorem, according to which, a^b will be transcendental if *a* is an algebraic number other than 0 or 1, and *b* is an irrational algebraic number. Using this theorem, we can deduce that *e* to the power of π (e^π) is transcendental since, if you recall, $e^\pi = e^{i\pi(-i)} = (-1)^{-i}$.

How? In this case, (−1), which appears in the place of *a*, is an algebraic number that is not 0 or 1. The value ($-i$) appears in the place of *b*, and it is indeed both algebraic (*i* is the root of $x^2 + 1 = 0$) and irrational.

This beautiful theorem was proven independently in 1934 and 1935 by Russian mathematician Alexander Gelfond and German mathematician Theodor Schneider. The Gelfond–Schneider theorem provides the answer to the second part of Hilbert's seventh problem in the list of the 23 unsolved mathematical problems that he presented to the International Congress of Mathematicians that convened in 1900 in the Sorbonne University in Paris.

Thus we now know that both $2^{\sqrt{2}}$ and $e^{\pi} = e^{i\pi(-i)} = (-1)^{-i}$ are transcendental numbers. Two down, an infinite number to go.

The continuum hypothesis and the missing axiom

At this point, we are aware that the cardinality of real numbers is greater than that of natural numbers. But *how much* greater? And why do we denote it as $\aleph$? Why not say that the cardinal number for real numbers is $\aleph_1$, which seems to be the obvious value after $\aleph_0$?

As we said before, the fact that a set is uncountable does not necessarily imply that it is of cardinality $\aleph$. This leads to the natural question: is there any set of numbers whose cardinality is greater than $\aleph_0$ yet smaller than $\aleph$? Georg Cantor asked this very question in 1877.

> *In mathematics, the art of proposing a question must be held of higher value than solving it.*
>
> (Georg Cantor)

Cantor's view was that no set existed with a cardinality that was greater than $\aleph_0$ or smaller than $\aleph$. In other words, he hypothesized that the cardinality of real numbers was $\aleph_1$. This hypothesis is known as the *continuum hypothesis*.

CONTINUUM HYPOTHESIS (CH)

There is no set whose cardinality is strictly between that of the integers, $\aleph_0$, and the real numbers, $\aleph$.

Over the course of many years, and despite great efforts, mathematicians have not been able to prove nor disprove this hypothesis. It was listed as the first problem of Hilbert's famous list of the 23 most important open problems in mathematics.

In order to understand the historical event which led to the resolution of CH, we must take a step back and observe what is happening to geometry at this time. If you recall, most of geometry is based on Euclid's system of axioms (aka postulates) that he developed over 2,000 years ago and which is still appropriate for what can be called "basic" geometry. Although ancient, there was a long-standing open problem regarding Euclid's fifth postulate, "the parallel lines axiom". According to this postulate, if there is on a plane, a line, *m*, and a point, A, not on the line, no more than one line parallel to the given line can be drawn through the point. (To be truthful, this is eighteenth-century Scottish mathematician John Playfair's version of Euclid's fifth postulate. Euclid's version involves sums of angles and does not mention the word "parallel".). The question was: can the fifth postulate be deduced from the other axioms? In other words, is this axiom redundant? It turned out that this axiom was essential, meaning that it couldn't be proven using only the other postulates. This idea probably inspired mathematicians to investigate how CH relates to the axioms of set theory, and, ultimately, the discourse regarding the axioms influenced set theory.

Over the years, it became clear that questions about infinity lay somewhere very close to the foundations of mathematics, and they should be treated with extreme caution.

In 1908, a set of axioms known as the Zermelo–Fraenkel system (ZF) was developed. We have already made Zermelo's acquaintance (the one who defended Cantor and formulated the first theorem in game theory);

Avraham Halevi Fraenkel is an Israeli mathematician who served as the first dean in the School of Mathematics in the Hebrew University of Jerusalem. They formulated this system in order to establish a solid ground for set theory and mathematics in general, with which mathematicians could formally address the concepts of cardinality of infinite sets, and settle some problems in this area, such as Russell's paradox. The ZF axioms are simply very elementary statements about the concept of sets that we believe (yes, in our hearts) are self-evident enough to be taken for granted. For example, here is the "empty set axiom":

$$\exists A \forall B (B \notin A)$$

Which is translated into human language as "there exists a set with no elements".

The Zermelo–Fraenkel axiomatic system was meant to serve the same role in set theory that Euclid's system of axioms serves in geometry. However, it was not quite meant to be.

In 1938, Austrian logician-mathematician-philosopher Kurt Gödel proved that it is impossible to *disprove* the continuum hypothesis using the Zermelo–Fraenkel axiomatic system of set theory. In 1963, 25 years later, Paul Cohen (1934–2007), a professor of mathematics at Stanford University, showed that it was impossible to *prove* the continuum hypothesis from the Zermelo–Fraenkel axioms. Cohen and Gödel had proven that the continuum hypothesis could neither be proved nor disproved. The result thereof is that the continuum hypothesis cannot be settled only by the ZF axioms. This was the birth of the first "undecidable" statement.

In the old Euclidean world, Aristotle's logic provided two cases of verity: true (T) and false (F). Now, we have acquired another value of verity: undecidable = U.

The obvious question one might ask is if this undecidable problem is due perhaps to there being some missing axioms in the Zermelo–Fraenkel system? It is still possible that there is another "obviously true" concept which has not yet been discovered and whose addition to the ZF system will allow us to prove CH. More optimistically, is it possible to amend ZF with some additional axioms so that every statement is decidable by it?

In 1931, Gödel, when he was only 25 years old, came up with three theorems: one completeness theorem and two incompleteness theorems, which talk about general undecidable statements. Basically, the first incompleteness theorem says that no matter with what axiom system one works, as long as it is sufficiently powerful to yield natural numbers, then an undecidable statement exists.[4] This was a surprising barrier on what one can expect of an axiomatic system.

These three theorems shocked the world of mathematics so much that, even today, people are still arguing over their essence. This fascinating subject is certainly one worth delving into for its own sake.

For more than 50 years, mathematicians in the field of axiomatic set theory have been trying to find the "missing axiom(s)". They have all been unsuccessful. Today, most experts in the field believe that there are no missing axioms and that the correct approach is rather to understand how different axioms relate to others. One could, of course, consider the continuum hypothesis to be an axiom, but it is important to remember that axioms must be phrased in such a way that their correctness can be easily believed, and this is certainly not the case with the continuum hypothesis.

In 2006 (one year before his death), Paul Cohen gave a fascinating lecture about the continuum hypothesis at the congress in honour of Gödel in Vienna. His lecture (in six parts) can be found on YouTube. (Search "Paul Cohen part 1 of 6, Centennial, Vienna.")

Meanwhile, in geometry, some interesting theories have risen out of the ashes with respect to the impossibility of substantiating the fifth postulate using Euclid's axiomatic system. In the nineteenth century, two more geometric systems were developed that are considered to be non-Euclidean geometry. The first one (hyperbolic geometry) assumes that *more than one* line that does *not* intersect a line *m* can be drawn through a point A that does not lie on line *m*. The second (elliptic geometry) assumes that it is *impossible to draw any* line that does *not* intersect line *m* through a point A that does not lie on a line *m*.

Similar to how new non-Euclidean theories have been introduced into geometry as a result of substantiating the fifth postulate in Euclidean geometry, the substantiation of the continuum hypothesis also led to the development of a non-Cantorian set theory in which the continuum hypothesis is not assumed. To be totally honest, there are many non-Cantorian theories, because mathematicians in recent years, by using Paul Cohen's systematic method of "forcing", have *proven the improvability* of many of the classic open unsolved problems.

In the past, it was acceptable to think that – provided mathematicians who were smart enough would work on them long enough – any mathematical claim could be either proven or refuted. Gödel's theorems prove that there are propositions that are not correct and are also not incorrect. They are, in fact, undecidable.

Mathematics may be defined as the subject in which we never know what we are talking about, nor whether what we are saying is true.

(Bertrand Russell)

Richard's paradox (about most things we are speechless)

The paradox we are about to discuss bears the name of French mathematician Jules Richard (1882–1956), and was published in 1905. Below I provide a verbal (as opposed to technical) description of the paradox.

The phrase "a real number, whose whole portion is 42 and whose digits after the decimal point are 0 in the odd places and 1 in the even places" precisely defines the number 42.0101010101... Likewise, the phrase "the number that if multiplied by itself three times will give us the number 7" precisely defines a number ($\sqrt[3]{7}$).

Richard said: Let us denote by E the set of all real numbers that can be defined by a finite number of words. This set is, without a doubt, countable (since we can arrange the numbers according to the quantity of words in the definitions, and for definitions of the same length, according to their lexographic (alphabetical) order. Next, using Cantor's diagonal method, he constructed a number which is not in the original set of numbers. Yet, the number can still be defined using a finite number of words. Thus, the number is *not* in the set, yet should be part of the set.

A paradox.

One way to settle this paradox is to point out that the property "a number that cannot be *defined* using a finite number of words" is not a property that can be defined in mathematical language. To expand this notion, let us observe this paradox from a different angle. We shall assume that the English dictionary comprises only five words, for example: "boy", "coy", "joy", "soy" and "toy". It is more than likely that with such a limitation, we will find it impossible to discuss any topic that requires the use of more than these five words. For example, we could not discuss the continuum hypothesis, and we certainly would not be able to discuss possible contradictions between different physical theories.

Every system of symbolic logic (mathematics included) comprises a collection of formulas. The use of the word "formula" here is not restricted to the relatively small mathematical sense of the word. It must be understood in a much broader sense: a symbol, word, expression, phrase, definition – anything that with its help we can express ideas. Since there is a 1:1 correspondence between the set of all formulas and the set of natural numbers, it is clear that the cardinality of the set of all the formulas is $\aleph_0$. If this is so, how can one discuss real numbers? Their cardinality is greater than $\aleph_0$. The conclusion is that there must be real numbers that cannot be described by formulas.

In this context, it is interesting to point out that the American mathematician and philosopher Charles Pierce, whom we mentioned earlier, also discovered, independently of Cantor, that it is impossible to find correspondence between the natural numbers and the real numbers. However, unlike Cantor, Pierce did not continue further. Instead, he decided that real numbers do not exist as a finished product and there is not much of importance that can be said about them.

Whereof one cannot speak, thereof one must be silent.

(Ludwig Wittgenstein)

Computable numbers

DEFINITION

A real number is a computable number if there exists some algorithm by which one can approximate the number's decimal development for any desired approximation.

Rational numbers are computable because their decimal development is finite, or infinite but repetitive and obtainable through the good, old-fashioned operation of division.

The number 0.232233222333222… is also computable, since its decimal development is easily found for any desired length. (Note: This number is *not* rational! Perhaps you would like to prove this.)

Algebraic numbers are also computable because there are various methods for solving any equation in the form $a_n x^n + a_{n-1} x^{n-1} + \ldots + a_1 x + a_0 = 0$ and finding its roots at any level of precision that we may desire.

And then, there are numbers that are none of the above, yet still computable. Two of these are π and e.

What is π?

The irrational number π has an infinite decimal development that never repeats and there is no algebraic formula for it. Yet, it is still computable.

Archimedes was already aware of an algorithm that could lead to a decimal solution for π with increasing accuracy. (This was based on circumscribed and regular polygons with n vertices inscribed in a circle. As n approaches infinity, the polygon approaches a circle in shape.)

In 1593, François Viète, a French mathematician, found a wonderful formula to compute π using a set of nesting radical numbers.[5]

$$\pi = 2 \times \frac{2}{\sqrt{2}} \times \frac{2}{\sqrt{2+\sqrt{2}}} \times \frac{2}{\sqrt{2+\sqrt{2+\sqrt{2}}}} \cdots$$

Besides the exceptional intrinsic beauty of this formula, it exhibits something very important, and that is the ellipsis at the end, which means "continue the process ad infinitum". You may not believe this, but this was the first time that an infinite process was clearly expressed in a mathematical formula.

This reminds me that it has been said that Ludwig Wittgenstein, in his lectures, asked his audience to imagine a man walking along and reciting "…, 5, 1, 4, 1, point, 3 – finished". When the man is asked what he is doing, he answers that he has just finished reciting the decimal expansion of π from the end to the beginning, a task he has been busy with for the past eternity. This story seems much more absurd than the one of a man who decides to sit down and write out the decimal solution of π from beginning to end, and will continue in this task for eternity. Why?

But back to π. It is interesting to note that, besides those of Archimedes and Viète, there were many other attempts to calculate the decimal value of π, all of which ultimately led to endless columns or endless multiplication operations. However, in 1650, English mathematician John Wallis discovered the following:

$$\frac{2}{1} \times \frac{2}{3} \times \frac{4}{3} \times \frac{4}{5} \times \frac{6}{5} \times \frac{6}{7} \times \frac{8}{7} \times \frac{8}{9} \cdots = \frac{\pi}{2}$$

By multiplying together consecutive pairs of multiplicands, this formula can be rewritten as follows:

$$\frac{4}{3} \times \frac{16}{15} \times \frac{36}{35} \times \frac{64}{63} \times \cdots = \frac{\pi}{2}$$

This infinite equation will, indeed, provide more and more digits in the infinite decimal development of π.

It is interesting to mention that it was John Wallis who in 1665 was the first to use the symbol of infinity ∞ (to tell the truth he used $1/\infty$ in his work about area calculations *De sectionibus conicis.*)

In 1671, Scottish mathematician and astronomer James Gregory came up with a different formula for π in the form of the infinite summation:

$$\frac{1}{1} - \frac{1}{3} + \frac{1}{5} - \frac{1}{7} + \frac{1}{9} - \frac{1}{11} \cdots = \frac{\pi}{4}$$

What a beautiful formula! Simple, elegant and impressive.

For the sake of proper disclosure, however, I would be amiss if I did not point out that, today, the credit for the discovery of the above formula is given to fourteenth-century Indian mathematician Madhava, who, it seems, was aware of it long before Gregory. Some scholars claim that not only was Madhava aware of the formula, he even found a way to calculate its deviation from the true value of π, and even devised another formula for π that approaches the value of π much more directly than that of Gregory. Here it is:

$$\pi = \sqrt{12}\left(1 - \frac{1}{3.3} + \frac{1}{5.3^2} + \frac{1}{7.3^3} + \cdots\right)$$

To be honest, I took advantage of this opportunity that came my way to show you some particularly beautiful formulas for calculating π. It would have been sufficient to show just one to prove that π is a computable number.

What is *e*?

Euler's number, *e*, is also not an algebraic number, but since it is defined as the limit of some sequence, its value is also computable and, similar to π, there are also a number of methods for computing its value. Here are a few nice, simple (relatively) examples. You may already be familiar with the first two.

$$e = \lim_{n \to \infty} \left(1 + \frac{1}{n}\right)^n$$

$$e = \sum_{n=0}^{\infty} \frac{1}{n!} = \frac{1}{0!} + \frac{1}{1!} + \frac{1}{2!} + \frac{1}{3!} + \frac{1}{4!} + \cdots$$

$$e = \lim_{n \to \infty} \frac{n}{\sqrt[n]{n!}}$$

For after all what is man in nature? A nothing in relation to infinity, all in relation to nothing, a central point between nothing and all and infinitely far from understanding either.

(Blaise Pascal)

Uncomputable real numbers

Are there any real numbers that are not computable? Not only are there, there are a great many. In fact, since, as we pointed out before, the number of algorithms are countable, the cardinality of computable numbers must be $\aleph_0$. And since the cardinality of all the real numbers is $\aleph$, this means that there must be $\aleph$ real numbers that are not computable! In other words, almost all the real numbers are not computable. There are no algorithms to define most of the real numbers. Can one talk about numbers that are uncomputable? Can you find an example of a real number that is not computable?

There are mathematicians who claim that there is no need to have the entire set of real numbers, and that, for all practical purposes, it is entirely feasible to make do with only the computable numbers.

For those who want to learn more (much more!) about computable numbers and their fascinating connection with the concepts of Alan Turing, I highly recommend reading *The Emperor's New Mind: Concerning Computers, Minds, and the Laws of Physics,* by British mathematician and philosopher, and the bearer of innumerable (might we say infinite?) awards and titles, Sir Roger Penrose.

IMPOSSIBLE SHAPES THAT GO ON FOR INFINITY

In collaboration with his father, Lionel Penrose, Sir Roger Penrose designed a number of impossible shapes and sent them to Dutch artist M C Escher (one of the heroes in *Gödel, Escher, Bach),* who used them in his etchings. The two most famous shapes are the following:[6]

The Penrose triangle

Penrose stairs – a never-ending journey

Just imagine climbing these stairs, and climbing and climbing yet always returning to the same spot. Escher added ever-climbing monks and ever-descending monks, all of whom end up in the same place they started.

Well, we have discovered a number of interesting concepts, but now the time has come for us to return to Cantor's theory and discover that infinity is infinite.

Infinity is infinite

Does a set of numbers exist whose cardinality is greater than that of the set of real numbers? Is there such a thing as a "maximum" value for cardinality?

Cantor himself demonstrated that there exists no set that has a maximum cardinality. In this case, Cantor's proof is actually constructive, because he shows how, for any given set, one can always find a set of even higher cardinality. This set has a very cool name, which is "power set".

Power sets

Before we get to the theorem itself, we shall learn a new concept.

> **DEFINITION: POWER SET**
>
> Assume a set A is given. The set that consists of *all the subsets* of A shall be termed the power set of A and is denoted by P(A).

For example, let us assume that A = {17, 42, 0}. We form the power set P(A) as follows: P(A) = {{}, {17}, {42}, {0}, {17, 42}, {17, 0}, {42, 0}, {17, 42, 0}}

The "{}" indicates the empty set, which is considered a subset of every set of A. (You may remember the use of the Ø symbol earlier in this book to denote an empty set. Both {} and Ø represent the same idea.) Note that set A is also considered a subset of itself. If you count the number of subsets, you will discover that there are three elements in set A, itself, and there are eight elements in the power set of A.

I'm sure that the equation $2^3 = 8$ immediately springs to mind. Is this relationship coincidental? No, it is not.

Mini theorem:
If #A = n, then #P(A) = 2^n. (# represents the number of elements.)

The proof for why the power set of a set with n elements will comprise 2^n subsets is sponsored by William Shakespeare, and goes like this: Every element in the original set must decide whether it is "to be or not to be" a member of any particular subset. Since every element has these two options for any particular subset, the total number of possibilities for n elements is therefore 2^n.

To explain this concept more concretely, assume that we are calling for members of a subset for A = {17, 42, 0}. 17 and 0 "decide" to be part of this subset whereas 42 declines. This combination of decisions results in the subset {17, 0}. The decision of each element unequivocally determines the make-up of a particular subset, and therefore the number of subsets is equal to the number of unique decisions, that is to say, $2 \times 2 \times 2 \ldots \times 2 = 2^n$.

QED.

Cantor's theorem
For every set A, its cardinality is strictly less than the cardinality of P(A).

To put this into layman's terms, Cantor's theorem means that the number of elements, #A, of a set A, will be strictly less than the "number" of subsets in its power set, P(A). That is to say, the power set of any set will have a greater cardinality than that of the set itself.

Now consider an infinite set, such as a countable set with cardinality $\aleph_0$ or a continuum set with cardinality $\aleph$. The cardinality of their power sets will be denoted as $2^{\aleph_0}$ and $2^{\aleph}$ respectively.

Two brain-twisters for mathematicians

1 Prove Cantor's theorem (hint: Russell's paradox).
2 Given that the cardinality of the set of natural numbers is $\aleph_0$, the cardinality of its subsets will be $2^{\aleph_0}$. Prove that $2^{\aleph_0} = \aleph$. In other words, prove that the cardinality of all the subsets of the natural numbers is equal to the continuum cardinality.

The Burali-Forti paradox

In 1897, Italian mathematician Cezare Burali-Forti presented a paradox that subsequently bore his name. It can be described as follows:

Let us examine the set of all the sets; that is to say, the set of all the people who are living today, the set of all the people who lived in the past, the set of all the songs that can be composed, the set of all the women who have never appeared on the Fashion Channel, the set of all women whose name is Grizelda, the set of all the flowers, the set of all the ideas that can ever be thought, the set of all the battles that I have not participated in, the set of all real numbers, the set of all the movies not directed by Tarkovsky, the set of all the movies that appear or have appeared on YouTube, the set of all the functions, the set of all the philosophers who never suffered from depression, the set of all the molecules that can be found at this moment in my basement … Now, add to that all the subsets of all these sets. In short, anything that you can possibly think of will be found in this set of sets.

Let us denote the set of sets by Ω.

It is clear that the cardinality of Ω must be greater than the cardinality of any of the other sets – because it includes everything. Yet, Cantor's theorem states that $\#P(\Omega) > \#(\Omega)$. That is to say, the cardinality of $P(\Omega)$ is greater than the cardinality of Ω, which is the set of all sets!

Cantor was not unduly disturbed by this paradox, because, in his opinion, the set of all sets was too large to be considered a set. Also, the reader should not be surprised, since in light of Russell's paradox, we know that not every collection of objects forms a legitimate set.

Cardinal arithmetic

I hope it is clear by now that the term "cardinality" is simply a generalization of the concept of "number of elements" used for finite sets, for infinite sets. The *natural cardinal* numbers are used to denote the number of items in a *finite* set, but *cardinals* also intuitively denote the number of items in an *infinite* set. For example, if a set has cardinality $\aleph_0$, then the quantity of elements in this set is the same quantity as in the set of natural numbers.

Now, we learned in mathematics class that finite numbers can be operated on mathematically using such operations as addition, division and multiplication. Furthermore, one can express these basic arithmetic operations as operations between sets. Consider that when we add two natural numbers together we are, in effect, "joining" them; this is parallel to joining two disjoint sets (disjoint sets are those that do not have any common members). If there are m members in one set and n members in the other, the union of the two sets will have $n + m$ items.

Here is a simple example:

If $A = \{Q, W, E, R, T, Y\}$ and $B = \{17, 21\}$, then $A \cup B = \{Q, W, E, R, T, Y, 17, 21\}$.

Here, $\#A = 6$ and $\#B = 2$, therefore $\#A \cup B = 6 + 2 = 8$.

Operations with cardinality work exactly the same. For example, in order to calculate $\aleph_0 + \aleph_0$ we must take two disjoint sets, both of which are countable, and see what the cardinality of their union set is. (You will see from the example that the result does not depend on the choice of sets.)

For example, we will take, $A = (1, 3, 5, 7, 9, 11, \ldots)$ and $B = (2, 4, 6, 8, 10, \ldots)$. A and B are disjoint, and the cardinality of each is, of course, $\aleph_0$.

As you can see, in this case, $A \cup B = N$, that is to say, their union gives us all the natural numbers, the cardinality of which we know to be $\aleph_0$.

The upshot here is that $\aleph_0 + \aleph_0 = \aleph_0$. (We really have not discovered anything new: we already know that the union of two countable sets is also a countable set.)

But caution is required here! Don't get carried away and think that you can apply all the regular mathematical rules to infinite values. For example, even though $\aleph_0 + \aleph_0 = \aleph_0$, we cannot subtract the term $\aleph_0$ from both sides since then we will end up with the absurd expression: $\aleph_0 = 0$, which is rather outlandish! So remember, when dealing with infinite values, a modicum of caution must be exhibited.

We can also express the multiplication operation as an operation on sets. When we multiply the natural number n by m, the result, in actual fact, is simply summing n to itself m times, i.e. $n + n + \ldots + n = n \bullet m$. Translating this to sets: given two sets A and B, we consider "B copies" of A, meaning that against every element b of B, we add a copy of A. For example, if A = {Q, W, E, R, T} and B = {17, 21, 33}, to multiply the two sets together, we join a "17" copy of A, a "21" copy of A, and a "33" copy of A. This can be denoted as:

$$\begin{aligned} A \times B = &\{< Q,17 >,< W,17 >,< E,17 >,< R,17 >,< T,17 >\} \cup \\ &\cup \{< Q,21 >,< W,21 >,< E,21 >,< R,21 >,< T,21 >\} \cup \\ &\cup \{< Q,33 >,< W,33 >,< E,33 >,< R,33 >,< T,33 >\} \end{aligned}$$

The set A × B has 15 elements, which is precisely the number of elements in A times the number of elements in B. But now, for an infinite example, we can claim that $\aleph_0 \bullet \aleph_0 = \aleph_0$. Again this is just our observation that the Hilbert hotel can accommodate countably many countable sets.

Nevertheless, if we fool around with arithmetic operations between infinite sets, we can discover a number of interesting results.

1 Since $\aleph_0 = \aleph_0 + \aleph_0$, then $\aleph_0 + n = \aleph_0$ for every finite number, n. This is because $\aleph_0 \leq \aleph_0 + n \leq \aleph_0 + \aleph_0 = \aleph_0$
2 If we take segment [0,1], whose cardinality we have shown to be $\aleph$, and add to it the segment (1,2], whose cardinality is also $\aleph$, we obtain the segment [0,2], whose cardinality, like all segments, is $\aleph$. Thus we get that $\aleph + \aleph = \aleph$. Note the

round parenthesis at the beginning of (1,2]. This indicates that "1" is not included in the set. The number 1 is removed so as to ensure that the two segments are disjoint.

3 We demonstrated that an infinite ray also has the cardinality of $\aleph$. One can view an infinite ray as the union of a countable set made up of infinitely many disjoint segments: [0,1], (1,2], (2,3], (3, 4], (4,5], …, and therefore $\aleph \cdot \aleph_0 = \aleph$.

4 If there is a curve that fills a square, then it follows that $\aleph \cdot \aleph = \aleph$. To see this, think of the square as a composition of horizontal lines. This means that a square is in fact $\aleph$ copies of lines, which is $\aleph$ copies of $\aleph$. Now a curve is simply a bent line, so its cardinality is $\aleph$. Since we can fill the square with the curve, we will in fact prove that $\aleph \cdot \aleph = \aleph$. The line segment [0,1] has the same cardinality as a square.

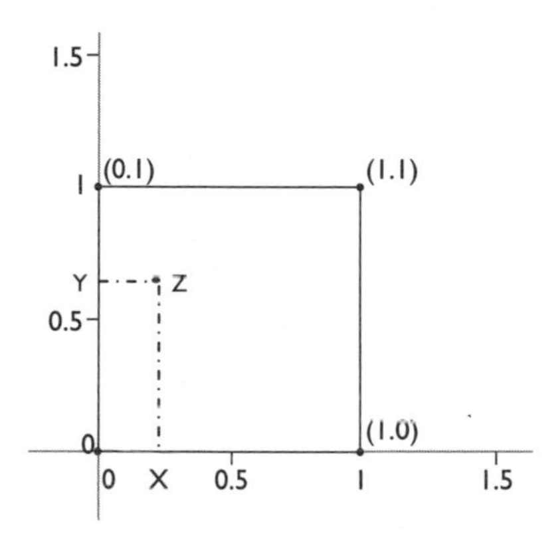

It is really not so difficult to directly prove that segment [0,1] has the same cardinality as the square. Let us examine the unit square above.

We choose any point in the square. Let us assume that the coordinates of our point are X = 0.a1a2a3a4… and Y = 0.b1b2b3…

We fit a point, Z, in segment [0,1] such that Z = 0.a1b1a 2b2a3b3... You can check that this is 1:1 and surjective.

Here is a little summary of our findings:

$\aleph = \aleph + n$
$\aleph = \aleph + \aleph_0$
$\aleph = \aleph + \aleph$
$\aleph = n \bullet \aleph$
$\aleph = \aleph_0 \bullet \aleph$
$\aleph = \aleph \bullet \aleph$
$\aleph = 2^{\aleph_0}$

In other words, all those cardinalities above are equal to each other!

And what is the significance of all this? Terms and equalities act very differently when we are in the world of the infinite.

The Cantor set

Another question Cantor asked was, is there an example of a set of cardinality $\aleph$ that does not contain a line segment? Indeed there is, and the example for such a set bears his name: the Cantor set. Its construction is as follows:

Let us divide line segment [0,1] into three equal parts and remove the middle segment, leaving only its endpoints.

We are left with segment [0,1/3] and segment [2/3,1]. Now, we proceed similarly, dividing these two segments into three equal portions and removing the middle one, except for its endpoints. We repeat this process (division into three and removing the middle portion) for each of the smaller portions we get after dividing [0,1/3] and [2/3,1], and so on and so forth for ever.

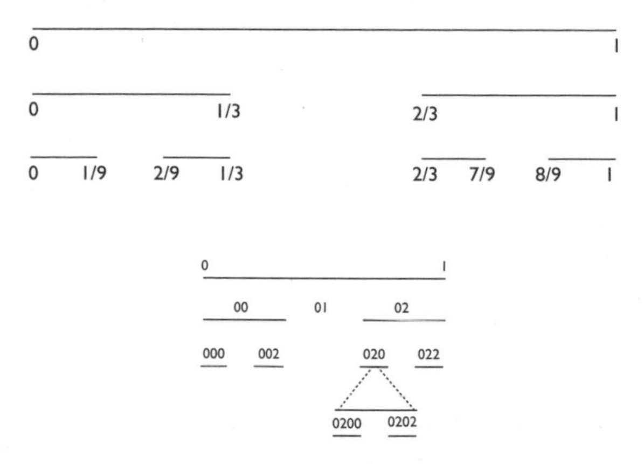

The set of all the points that appear in all the sets along the process after repeating it infinitely many times is called the "Cantor set". For example, 0 is a member of the Cantor set. This set has many interesting properties relating to topology, measurement, geometry, as well as set theory.

The Cantor set has a more concrete description. Because each segment is divided into three portions each time, it is convenient to use ternary representation (that is, using only the digits 0, 1, 2). It is not at all difficult to write numbers using ternary representation.

The ternary representation of a number, a, is $a = 0.c1c2c3\ldots$ where $a = c1/3 + c2/9 + c3/27 + \ldots$ The values 3, 9, 27 replace 10, 100, 1,000 of the more common decimal method.

For example:

$0.3 = 3/10 = \underline{0.0220022002200220}\ldots$

$0.5 = 5/10 = 1/2 = \underline{0.1111111}\ldots$

$0.8 = 4/5 = \underline{0.2101210121012101}\ldots$

(The line under the digits indicate that we are using ternary development.)

Thomas Aquinas (1224–1274) stated that even if infinite days pass, not one of them will be at an infinite amount of time from this moment. Likewise, on a real, infinite line, it holds that the distance between any two points will always be finite and, as Hegel said, infinity cannot be found anywhere on the infinite line.

The concrete description of the Cantor set is given in the following challenge.

An easy brain-twister

Show that all the points in the Cantor set are those in which the digit 1 does not appear in their ternary development.

It is now easy to see that the Cantor set has cardinality $\aleph$ because in the Cantor set are only numbers that use only the digits 0 and 2 when written in ternary. Nevertheless, it is clear that this set of numbers has the same cardinality as the set of numbers that can be written with only 0 and 1 digits. Writing numbers using only 0 and 1 digits is simply the binary method of writing numbers, and the method can be used to write all the numbers between 0 and 1. Therefore, the conclusion is that the Cantor set has the same cardinality as the set of all the numbers in the segment [0,1] and therefore its cardinality is $\aleph$.

This fact is quite surprising since the Cantor set has no length whatsoever. This is because the sum of the lengths of the segments we removed from it is:

$$1 \bullet \frac{1}{3} + 2 \bullet \frac{1}{9} + 4 \bullet \frac{1}{327} + 1 \bullet \frac{1}{3} + 8 \bullet \frac{1}{81} + \cdots = \frac{\frac{1}{3}}{1 - \frac{2}{3}} = 1$$

So the Cantor set length is the outcome of subtracting from 1 the length of all those segments, namely 1, which indicates that the length of the Cantor set is 0.

The Cantor set is indeed very special. It has an uncountable number of points – the total length of which is zero! – that can be found on a set of line segments! The Cantor set is also considered the oldest fractal. But this subject will need to wait for another book.

A LITTLE BIT MORE ABOUT REPRESENTING NUMBERS

By the way, the number 1 can be written in ternary notation as 0.2222…, and in decimal notation as 0.9999999… Many people raise their eyebrow (or even both of them) when I write that 1 = 0.999999… They try to explain to me that this is incorrect and that 1 is ever-so-slightly greater than 0.999999…

For the most part, it is almost impossible to convince people of my correctness. But that doesn't mean that I won't try.

Try to subtract 0.9999…. from 1. What do you get? If your result is any different from zero, you are using incorrect logic.

Or, try this: Let $a = 0.9999999\ldots$ In that case, $10a = 9.999999\ldots$ If we subtract one from the other, $10a - a = 9.999999\ldots - 0.999999\ldots$ This very nicely leads to $9a = 9$, or $a = 1$.

And if that hasn't convinced you, I'm really sorry.

CONCLUSION

A book about infinity cannot have an end; infinity is a never-ending story. So, instead of writing a conclusion, I will present to you a very beautiful problem and you will think about it as long as you wish…

Take a look at the following:

1/9,801=
0.0001020304050607080910111213141516171819202122232425262728293031323334353637383940414243444546474849505152535455565758596061626364656667686970717273747576777879808182838485868788899091929394959697990001 0203…979900010203…

Do you see what is happening here?

No?

OK.

Now you have it in a better resolution:

1/9,801=
0.00 01 02 03 04 05 06 07 08 09 10 11 12 13 14 15 16 17 18 19 20 21 22 23 24 25 26 27 28 29 30 31 32 33 34 35 36 37 38 39 40 41 42 43 44 45 46 47 48 49 50 51 52 53 54 55 56 57 58 59 60 61 62 63 64 65 66 67 68 69 70 71 72 73 74 75 76 77 78 79 80 81 82 83 84 85 86 87 88 89 90 91 92 93 94 95 96 97 99 00 01 02 03…97 99 00 01 02 03 04 05 06… ad infinitum.

We have here all the two-digit numbers in perfect order (!) and repeating themselves to infinity, but the number 98 is missing.

Brain-twister

Why is 98 missing?

Hint: is 98 really missing?

What will happen if we try 1/1,089?

What will happen if we try 1/ 998,001?

I'll finish the text with my favourite word:

WHY?

NOTES

Warm-up

1 You can look it up by Googling "Elvis Presley Kevin Bacon". Elvis Presley was in *Change of Habit* (1969) with Edward Asner. Edward Asner was in *JFK* (1991) with Kevin Bacon, therefore Asner has a Bacon number of 1, and Presley (who never appeared in a film with Bacon) has a Bacon number of 2.

2 Go is an abstract strategy board game for two players, in which the aim is to surround more territory than the opponent and requires the skills of strategy, tactics and observation. The goal of Gomoku (also called Gobang or Five in a Row), while also an abstract strategy board game and traditionally played with Go pieces on a 15 × 15 or 19 × 19 Go board, is for a player to be the first to form a line of five markers. It can also be played using pencil and paper.

3 I first saw this riddle of the climbing monk in Martin Gardner's *My Best Mathematical and Logical Puzzles*. This small book is extremely entertaining.

4 Many mathematicians will disagree with this. They will say that we are talking here about limits of convergence, and it all depends on the type of convergence involved. Readers who are not mathematicians might like to look up the concept "Supertask" on Wikipedia: carrying out infinitely many tasks in a finite span of time. We will encounter this concept later when we meet Zeno, Achilles, and a tortoise.

Lesson 1

1 Ironically, hardly anything is known about Laërtius's own life; all we know is that the great biographer lived "sometime in the third century".
2 Quoted from *The History of Western Philosophy* by Bertrand Russell.
3 The condition is quite technical, so I will not elaborate.
4 Thābit Ibn Qurra was also one of the first who expanded Pythagoras's theorem for right-angled triangles into one that applied to all triangles.
5 The term "proper" means that the set of divisors does not include the number itself.

Lesson 2

1 Harald Bohr, a mathematician, was the great Danish physicist Niels Bohr's brother. He was also a member of the Danish national soccer team, with whom he won an Olympic silver medal in 1908. Quote from his "Looking Backward", *Collected Mathematical Works*.
2 Fujiwara is well-known in Japan as the author of popular mathematics books. One of his books is about the beauty of theorems, where he categorized them as beautiful or ugly.
3 The first number is 81. The second is … drumroll! 1,458. Did you find it?
4 The answer is 62. Each number is the sum of the previous number and the sum of the previous number's digits. For example, the number that follows 16 is 23 since 16 + (1 + 6) = 16 + 7 = 23. The answer to the riddle, therefore, is 49 + 13 = 62.

Lesson 3

1 Hint: Find the maximum power of 2 that divides your number. This should be $p - 1$.
2 The prime numbers theorem states (more or less) that the chance that a number close to n will be a prime number is proportional to the logarithm of n divided by n. Since this proportion tends to 0 as n approaches infinity, this guarantees the sparseness of the prime numbers among all natural numbers.
3 A prime triplet is a set of three prime numbers of the form $(p, p + 2, p + 6)$ or $(p, p + 4, p + 6)$. This is the closest possible grouping of three prime numbers, since one of every three sequential odd numbers is a multiple of three, and hence not prime (except for 3 itself), except (2, 3, 5) and (3, 5, 7).
4 100! is, to be precise:
93,326,215,443,944,152,681,699,238,856,266,700,490,
715,968,264,381,621,468,592,963,895,217,599,993,
229,915,608,941,463,976,156,518,286,253,697,920,
827,223,758,251,185,210,916,864,000,000,000,000,
000,000,000,000.
5 This name is the result of a mistranslation. The original name was *versiera* from the Italian, which means "to turn". However, the word can also be viewed as an abbreviation for *avversiera*, which means "adversary" and is commonly used to refer to a female devil. When the name of the curve was translated into English, somebody (either on purpose or innocently) used the word "witch" (which is pretty much a female devil).
6 They are named after her because she used them in her investigations of Fermat's last theorem.

Lesson 4

1 The "Theaetetus" is one of Plato's dialogues concerning the nature of knowledge, written circa 369 BC.
2 In addition to his discovery of irrational numbers, Pythagoras made another important contribution to the science of mathematics, and that was the introduction of the concept of "proof" in a sense very similar to what we know today.
3 In case you have forgotten, the logarithm is the inverse function to exponentiation. That is to say, if $b^y = x$, then $\log_b(x) = y$. In other words, the logarithm of a given number x is the exponent to which another fixed number, the base b, must be raised, to produce that number x. For example: 1,000 is 10^3, so $\log_{10}(1{,}000) = 3$. Similarly, $\log_2(64) = 6$ since $2^6 = 64$.
4 Two quantities are in the golden ratio if their ratio is the same as the ratio of their sum to the larger of the two quantities (i.e. if $a>b$, then if $a/b = (a+b)/a$, a and b are in the golden ratio). The golden ratio is represented by φ.

Lesson 5

1 On 23 March 2010, the charitable organization The Warm House, in an open letter on the internet, asked Perelman to donate the money to them. The Clay Institute subsequently used Perelman's prize money to fund the "Poincaré Chair", a temporary position for young, promising mathematicians at the Paris Institut Henri Poincaré.
2 I found this example in Jose Benardete's book, *Infinity*.

Lesson 6

1 For anyone interested in the biography of Georg Cantor, I highly recommend the excellent book by Dauben, *Georg Cantor: His Mathematics and Philosophy of the Infinite.*

2 Zermelo is also responsible for an important theory in chess, which some see as the first theorem in game theory. (I explained about it in my book, *Gladiators, Pirates and Games of Trust*, Watkins, 2017.)

3 Russell is also the recipient of a Sylvester Medal. He also won the Nobel Prize in Literature in 1958. As far as I know, Russell is the only individual ever to win both these prestigious prizes.

Lesson 7

1 When the height is equal, the numbers are ordered according to their numerators.

Lesson 8

1 I once read a book in Russian about set theory that was published before Perestroika, and it included Hebrew letters. If that doesn't impress you, I will just add that in those days they would jail people for learning Hebrew.

2 For mathematicians: Algebraic numbers are closed under the operations of multiplication and addition, that is to say, they form a "ring".

3 Euler's constant, *e,* is the "star" of the most beautiful equation in the world: $e^{i\pi} + 1 = 0$. The number, 2.17181828459045…, pops up in a number of places in

mathematics. It is used as the base for natural logarithms, in natural exponential functions, in calculating interest, and more.

4 There are other natural requirements of this system that ZF satisfies.

5 A radical is simply any number that has the "root" symbol in it.

6 Penrose, L S & Penrose, R. "Impossible Objects: A Special Type of Visual Illusion." *British Journal of Psychology* 49 (1958): 31–33.

FURTHER READING

If you are interested in studying the subject more, following is *a very short list* of some of the books that I think are particularly worth reading.

The Music of the Primes by Marcus du Sautoy
One Two Three… Infinity by George Gamow
The Colossal Book of Short Puzzles and Problems by Martin Gardner
Satan, Cantor and Infinity by Raymond Smullyan
Gödel, Escher, Bach by Douglas Hofstadter
A Mathematician's Apology by G H Hardy